Palgrave Frontiers in Philosophy of Religion

Series Editors
Yujin Nagasawa
Department of Philosophy
University of Birmingham
Birmingham, UK

Erik J. Wielenberg
Department of Philosophy
DePauw University
Greencastle, IN, USA

Palgrave Frontiers in Philosophy of Religion is a long overdue series which will provide a unique platform for the advancement of research in this area. Each book in the series aims to progress a debate in the philosophy of religion by (i) offering a novel argument to establish a strikingly original thesis, or (ii) approaching an ongoing dispute from a radically new point of view. Each title in the series contributes to this aim by utilising recent developments in empirical sciences or cutting-edge research in foundational areas of philosophy (such as metaphysics, epistemology and ethics).

More information about this series at
http://www.palgrave.com/gp/series/14700

Daniel Saudek

Change, the Arrow of Time, and Divine Eternity in Light of Relativity Theory

palgrave
macmillan

Daniel Saudek
Philosophisch-Theologische
Hochschule Sankt Georgen
Frankfurt am Main, Germany

Palgrave Frontiers in Philosophy of Religion
ISBN 978-3-030-38410-4 ISBN 978-3-030-38411-1 (eBook)
https://doi.org/10.1007/978-3-030-38411-1

Cover illustration: jvphoto / Alamy Stock Photo

This Palgrave Macmillan imprint is published by the registered company Springer Nature Switzerland AG.
The registered company address is: Gewerbestrasse 11, 6330 Cham, Switzerland

Acknowledgements

Substantial parts of this book are the result of the research which I undertook while employed at the Department of Christian Philosophy of the University of Innsbruck, Austria, between 2011 and 2016. I wish to thank Winfried Löffler for many patient discussions and much helpful feedback in those years, as well as Alberto Strumia for detailed feedback especially on the physics and mathematics in this book. Generous financial support enabling this research came from the John Templeton Foundation, the University of Innsbruck and the *Tiroler Wissenschaftsfonds*. Then, in 2016, I became a teacher at the *Vinzentinum* in Brixen/Bressanone, northern Italy, and in addition began teaching at the universities of Konstanz and Sankt Georgen, both in Germany. There was little time for research in the philosophy of physics in the years which followed, but many an inspiration came from the wonderful students—both at school and at university—whom I had the fortune to teach, with their bright, inquiring minds. Long train rides between Italy and Germany offered a bit of time needed for the hard thinking which led to some of the key results in this book. My continuing involvement with official academia, writing a few articles and attending the odd conference, alongside my main job as a teacher, would not have been possible without the constant support of the headmaster of the *Vinzentinum*, Christoph Stragenegg. Finally, the last parts of this book were written in the cheerful, always humorous atmosphere of the oratory of Frankfurt, where I moved in 2019 to begin working in the campus work of the diocese of Limburg. The editorial team of Palgrave Macmillan provided helpful, uncomplicated support in the finishing phase of this book.

If two people are mentioned last, it is because their presence accompanied the whole story of this book: Georg Gasser, who always had time for talking about my research and who taught me to navigate the tricky waters of the often confusing, scientometrically dominated world of contemporary scientific publishing. And my father, with whom I had many long conversations on time, space and causation, which served as a constant stimulus to sharpen the arguments in this book and to write with the greatest clarity and rigour possible.

Contents

List of Figures

1

Introduction

The aim of this book is twofold: in the first part, it is to propose a new solution to the problem of how relativity theory relates to the passage of time; in the second part, it is to spell out, in the light of this solution, the relationship between divine eternity, change, and time.

The issue of the first part of the book is one which the philosophy of time has been grappling with for over a century: On the one hand, time appears to be fundamentally asymmetric in character, being marked by a unidirectional flow from a fixed and irretrievable past towards a future which is open and can be influenced. Moreover, there is an overwhelming impression that past, present and future events differ fundamentally in their ontological status. On the other hand, strong arguments deriving from physics, especially relativity theory, speak against the above common-sense view of time. In particular, the notion that there is such a thing as a state of the universe at a certain time, and hence also a present state of the universe which separates its past from its future, are strongly called into question, so that past, present, and future events must be given the same ontological status. This point is reinforced by philosophical arguments which show the notion of a passage of time to be deeply problematic and even aporetic.

D. Saudek, *Change, the Arrow of Time, and Divine Eternity in Light of Relativity Theory*, Palgrave Frontiers in Philosophy of Religion,
https://doi.org/10.1007/978-3-030-38411-1_1

A further crucial datum which philosophy of time must take into account in this context is the fact that contemporary science describes a universe which undergoes a development from earlier to later stages, marked by different physical characteristics, from the big bang to the stage we are living in. Yet how is such an evolutionary cosmology possible if there is no such thing as the universe at a certain time?[1]

Philosophy of time therefore faces the task of "saving the phenomena" and accounting for why time seems as it does, and furthermore of assessing whether and how a truly evolutionary cosmology is possible, given the strength of the arguments against a common-sense view of time. Although much has been written on this subject, philosophy is still struggling with it. In my view, there is a too strong polarization between two opposing views, with still little on offer by way of a third alternative: On the one hand, there is the block universe model, whereby the abolition of a global past-present-future distinction results in a universe in which all events are in a certain sense already there and whose history is "written" and laid down, whereas our intuitive view of a flowing time, and our impression that the future contains several possibilities, is illusory, a position which has also become known as eternalism. On the other, the defenders of the intuitive view of time, known as A-theorists, have by and large subscribed to an absolute passage of time, with a global moving "now" and an associated past-present-future distinction valid for the entire universe, believing this to be necessary for there to be true becoming and a genuinely evolving world, and often positing quite implausible physics in order to make such a model possible.

To do justice to a changing and evolving world which, as I believe is convincingly shown by relativity theory, lacks a global past-present-future distinction, a new model is needed. The one developed in this book draws on arguments and insights coming from both physics and philosophy. Crucially, it is made possible by a theory whereby time is due to change in things, so that time is not itself a thing, or substance, existing in its own right. In particular, time is not thought of as a succession of instants,

[1] Contemporary cosmology is "evolutionary" in the sense that it describes a universe in systematic change and attempts to explain its development, which does not necessarily mean that the explanatory paradigm of biological evolution—random mutation plus natural selection—carries over to cosmology.

or "nows", independent of change occurring in the world. I will therefore make much use of Aristotelian and scholastic theory of time which, in my view, has a lot to offer for the task of spelling out the relationship between change and time, as well as insights from set theory and mathematical analysis.

Such a model, as I will argue, is up to the task of explaining phenomena such as time's one-way street character and the fact that agents can influence their future, but not their past. It furthermore allows for true becoming, and in combination with results from general relativity theory, it also makes a genuinely evolutionary cosmology possible. However, the theory proposed here is not an A-theory: it does without a global tide of becoming and without the notion that events change their ontological status by flowing from the future via the present into the past. Rather, all really occurring events, considered in themselves, have the same ontological status.

The model proposed in this book builds upon much groundwork in both physics and philosophy which must be considered first. Chapter 2 provides an overview of the available models of time concerning questions such as the following: Is time a thing, or substance, in its own right? Is there a true passage of time and an associated distinction between past, present, and future, or must this common-sense view of time, as many have argued, be rejected as illusory? The chapter places particular focus on the temporal ordering of events, and its final section is devoted to the notions of the instant moment and that of the universe at a given time.

Special relativity theory has decisively influenced the philosophy of time. Chapter 3 therefore discusses, first, the immediate consequences of this theory: the loss of absolute simultaneity, but also the invariant temporal order of causally connectable events. Next, the arguments advanced by Wim Rietdijk and Hilary Putnam in favour of universal determinism which, as they claim, results from relativity of simultaneity, is discussed, as well as the critique of this argument. It will also emerge that proposals for locating an observer-independently significant past-present-future distinction in Minkowski spacetime lack credibility. The chapter closes with a point of philosophical importance emphasized by Hans Reichenbach: the structure of world lines and light cones in relativity only works for relatively large, composite objects, and breaks down at the quantum level.

Contemporary evolutionary cosmology describes a universe whose different overall properties change with its age, a finding which seems to contrast with the upshot from Chap. 3 and to restore the notion of the universe at a given time. As will be argued in Chap. 4, accounts of the development of the universe with time can be given in a unique way by using the so-called cosmic time functions which can be defined in general relativity theory, given that we inhabit a homogenous and isotropic universe. After introducing cosmic time and its role in cosmology, I will explore its scope and limits. It will become clear that cosmic time does not save a global past-present-future distinction or a total temporal ordering of all events. The final section of Chap. 4 assesses the argument whereby quantum mechanics establishes an absolute, observer-independent simultaneity between events, and hence also a true "now". We will see, however, that this argument backfires.

The apparent contrast between the obviously asymmetric characteristics of time and the results of spacetime physics has generated a substantial body of literature addressing this problem. Chapter 5 gives an overview and assessment of an—obviously not exhaustive—choice of relevant works. It illustrates the polarization between the two opposing camps mentioned above, but also introduces the work of some authors who have proposed a third way, arguing for true becoming, while rejecting the global advancing "now".

This latter concept, as I argue in Chap. 6, does not help us to construct a physically and philosophically plausible model of time: the global moving "now" not only conflicts with relativity theory, but also it has no causal role, explains nothing, and only creates needless philosophical problems. Nor, as will be seen, are irreversible processes in physics good candidates for explaining the one-way street character of time.

Chapter 7 argues that the way forward to meet the challenge stated initially lies in a theory which understands time as based on causal interactions between substances, one which entirely dispenses with the global moving "now". In order to develop such a theory, I first consider aspects of Aristotelian and scholastic thought on the topics of time and change, as developed by Aristotle himself and by Francisco Suárez, but also useful work by Augustine on the measurement of time and the temporal asymmetry experienced by agents. This points towards a view of time as "spanned" by objects possessing contradictory properties. There are, however, many difficult objections against a theory based on causal interactions being able to account for the familiar features of time: How come that the same type of change can occur faster or more slowly, if there are only causal interactions, but no time as an entity in its own right? Can such a theory explain why causal processes take time in the first place? What about the one-way street character of time as experienced by conscious agents, and the asymmetry between the fixed past and the open future? At the most general level, how can causal interactions and change occur unless there is an underlying passage of time in which they do so? These questions set the agenda for the following chapters, in which a local theory of time is systematically developed.

Chapter 8 examines the issue of the duration of causal interactions: It is shown that the same type of causal process can last longer or shorter without presupposing an absolute background time. Also, it is argued that causation is better thought of as simultaneous rather than diachronic, but that nevertheless causal processes can take a nonzero time to unfold. In the context of this discussion, the medieval concept of the *aevum* is resuscitated as a heuristic help to understand the relationship between change, duration, and time.

Next, in Chap. 9, a full derivation of the arrow of time as a purely local concept, in no need of a global passage of time, is given on the basis of simple axioms. From atemporal notions, a definition of "before" is developed, whereupon definitions of change, and then of time as the quantification of change, are derived. It is then proved that time has a linear structure, and that the notion of changing the past entails a contradiction in a way in which that of changing the future does not. Local time thus turns out to exhibit the familiar structure of a one-way street with a fixed past and an open future. Time's passage is real, but local.

In Chap. 10, a series of objections which can be made against the theory of local time is discussed. Also, the theory is compared to other operationalist theories of time, as well as to causal ones. Chapter 11 then lays out the general traits of the worldview which emerges from the investigations in all the previous chapters: The world lacks a global past-present-future distinction, yet the worldlines of local agents are truly divided into a fixed past and an open future, so that time as we experience it is not illusory. We live in a fully consistent world where causal stories remain intact and nobody can travel backwards in time to, say, kill their ancestors. Nevertheless, a Gödel universe—one with closed timelike curves, as mapped by a coordinate observer—remains, as is shown, a possibility.

In the shorter second part of the book, the results attained in Part I are applied to the relationship between God, change, and time. First, in Chap. 12 it is argued that God as the ground of being can only be perfect and unlimited in all respects. From this, a simple argument shows that God cannot be subject to change. But if this is so, the question arises, how is it possible for God to act, either to create or to interact with the world? The reasons why limited beings necessarily change in their interactions do not, as will be shown, apply to God, so that, surprisingly perhaps, God can act without changing. Then, Chap. 13 discusses the relations which hold between God and temporal being: God's knowledge of creatures and their temporal perspective is in no way limited by His not being in time. In addition, the classical problem of divine foreknowledge and free will, which has likely caused many a headache since antiquity, is shown to simply dissolve in the light of relativity theory. The same is true of the problem raised by Norman Kretzmann of whether God can know what time it is. As for the local past, since the notion of changing it is, from Chap. 9, contradictory, so is that of God changing it. Chapter 14, finally, provides a rough, concluding outline of the drift of the book by zooming in from unlimited and eternal being, via successive limitations, to time as we know it. It turns out that Plato's adage whereby time is a "moving image of eternity" has much to say even today.

Since physicists and philosophers often have a hard time understanding each others' languages, I have attempted to make the material accessible to both, and indeed to any readers interested in the problem of time, employing wherever possible plain English, rather than formalism, and supplementing the necessary formalism with examples. Appendix A gives a simple, qualitative outline and explanation of elements of the special theory of relativity which are used in this book. Appendix B presents some key results and equations of general relativity, with sundry references to sources providing the derivations of these results for the interested reader. I have also translated quotations from non-English sources (mainly Latin and German ones), or, where indicated, employed already existing translations. All internet references in this book were still retrievable as of August 2019, unless a different date of retrieval is given.

Part I

Relativity Theory and the Passage of Time

2

Models of Time: A Brief Survey

It will first be necessary to provide a systematic overview of possible concepts of time which have been elaborated throughout the history of thought and in contemporary literature. The main points of contention have concerned time's ontological status and its structure.

2.1 Models of Time's Ontological Status

Is time an entity in its own right? That is, does its existence depend on the existence of other entities, or would it exist even in their absence? The former view is known in the literature as "relationism", whereas "substantivalism" has become the accepted term for the latter. Thus, on relationism, time is not a thing, but is instead constituted by relations between things. Substantivalism, on the other hand, holds that a complete list of existing things must include time.[1]

The substantivalist-relationist controversy must be distinguished from the related but separate issue of whether time is a real, objective feature of

---

[1] B. Dainton, *Time and Space*, 2nd edn (2010), pp. 1–3.

D. Saudek, *Change, the Arrow of Time, and Divine Eternity in Light of Relativity Theory*, Palgrave Frontiers in Philosophy of Religion,
https://doi.org/10.1007/978-3-030-38411-1_2

the world or a subjective and mind-dependent one. Whereas on substantivalism, time is objectively real, conversely relationism does not entail that it is necessarily a purely subjective phenomenon. Rather, its objectivity depends on whether the relations in question are themselves mind-independent, objective entities.[2]

The question of whether time is a "substance", and whether it "exists in its own right", suffers from a certain ambiguity inherent in these notions. By a "substance" one can mean, first, an irreducible element of reality, something whose existence cannot be explained in terms of something else. Second, the term can mean something like "object": objects can be said to exist in their own right, as opposed to, for example, their qualitative or quantitative properties, their boundaries or the relations between objects, all of which cannot exist without the objects themselves.[3] But that, of course, is not necessarily to say that objects such as cats and stars are irreducible elements of reality. Since time, on the other hand, seems to be pervasive and all-encompassing, and since it is invisible and does not appear to be an object, that is, a substance in the second sense of the word, substantivalists usually hold that time is a substance in the stronger sense of something ontologically basic.

Time's ontological status has been subject to controversy throughout the history of philosophy and science. This issue has consistently been discussed together with ontological problems surrounding change, motion, causality, and space, which are inextricably linked to it. The substantivalist-relationist controversy about time has a counterpart in the analogous debate about space. This, of course, is not to say that one's position in either debate necessarily commits one to holding the analogous position in the other. Relationism concerning time is associated with thinkers such as Aristotle and a number of medieval and early modern philosophers such as Thomas Aquinas, Suárez, and Leibniz. Substantivalism is perhaps most famously linked to the name of Isaac Newton, according to whom "absolute, true and mathematical time, in itself and of its own

---

[2] Clearly, relations come in both varieties: some are mind-independent (e.g. "greater than"), whereas others are not (e.g. "frightening to").

[3] Cf. H. Robinson, "Substance", in E. N. Zalta (ed.) *The Stanford Encyclopedia of Philosophy* (Spring 2014 Edition).

nature, flows equably without relation to anything external".[4] The notion of a substantival and absolute passage of time is also defended by quite a number of contemporary thinkers, as we will see in Chap. 5. The substantivalism-relationism debate has become epitomized in the correspondence between Gottfried Leibniz and Samuel Clarke. While this has centred mainly on substantivalism versus relationism concerning space, it contains much of relevance also for the debate on time. Clarke defended Newton's view of absolute space and time, whereas Leibniz vigorously attacked this position, basing his arguments in large parts on the principle of the identity of indiscernibles. Theological considerations are crucial to the debate between substantivalism and relationism, because a view of space and time as ontologically basic raises the question of how God, as the ultimate foundation of all that exists, is related to these realities. This is illustrated by the fact that Newton, influenced by religious currents of his time, identified space—albeit in somewhat vague terms—with a *sensorium* of God, whereas Leibniz warned that such a move leads to pantheism.[5] Similarly, the assumption of an absolute time leads to several problems concerning God's relation to it, problems which will be explored in more detail especially in Part II of this book.

Einstein's special and general theories of relativity decisively influenced the direction of the substantivalist-relationist debate. While these theories forgo a concept of time as being "without relation to anything external", they are compatible with an interpretation which views time and space as substances influenced by the matter embedded in them, but existing independently of it. Indeed, the view of Einstein himself can probably be characterized in this way, although there are certain ambiguities on this point in his writings.[6]

---

[4] "Tempus Absolutum, verum et mathematicum, in se et naturâ suâ sine relatione ad externum quodvis, aequabiliter fluit, alioque nomine dicitur Duratio." I. Newton, *Philosophiae Naturalis Principia Mathematica*, (1739), p. 12.

[5] See, for example, Dainton (2010), pp. 175–181; W. Pannenberg, "Eternity, Time and Space", *Zygon* (2005). On the meaning of the *sensorium* in Newton's thought, cf. S. Ducheyne, "Isaac Newton on space and time: metaphysician or not?", *Philosophica* (2001), especially pp. 88–93.

[6] For a discussion, see G. Schwarz, *Raum und Zeit als naturphilosophisches Problem*, 2nd edn (1992), pp. 191–194.

2.2 Models of Time's Structure

On the phenomenological level, time has some very peculiar features which set it apart from space and which have never ceased to puzzle philosophers and scientists. First, its characteristic flow or passage, whereby events come and subsequently seem to cease to exist. Then, its unidirectionality, whereby time always travels "forward" and, unlike points in space, no point in time can be revisited twice.[7] And finally, the distinction between the "past", "present", and "future" regions of the timeline, which appear as fundamentally different: The past seems "fixed" or "determinate"[8] but somehow unreal, or at any rate "less real" than the present. The future seems not to have occurred yet, and on this account lacks the determinate and concrete character of the past, so that it might be likened to a blurred image. The fleeting present seems to be the region of infinitesimal duration where past and future meet, and where open possibilities become concrete actualities, although these latter, "fully real" for an instant, are immediately swept away into the past. These features of time give rise to the impression of a "change in the existing"[9]: the set of existing events or facts is different at different points in time and not everything exists simultaneously.

2.2.1 General Characteristics of A- and B-Models of Time

But are these phenomenological characteristics of time, so strange and yet so familiar, real or merely apparent? On the one hand, they are experienced as ubiquitous and pervasive features of reality, which makes them hard to deny. On the other, the puzzling, even aporetic character of time as we experience it suggests that the "naïve" view of time cannot be upheld

[7] Cf. E. Runggaldier, C. Kanzian, *Grundprobleme der Analytischen Ontologie* (1998), pp. 100–104.

[8] The term "determinate" will frequently be used in the following discussion. The precise meaning of this term is left open for the sake of a broad characterization of current models of time. A formal definition will be given in Sect. 2.2.3.

[9] See the discussion in M. Dorato "On becoming, cosmic time and rotating universes", in C. Callender (ed.) *Time, Reality and Experience* (2002), here especially pp. 260–265.

and must be revised, or even rejected outright. In contemporary analytic philosophy, those who hold that flowing time is real are known as A-theorists. B-theorists, on the other hand, deny time a flowing character and liken the structure of time to that of space. The by now classical distinction between A- and B-theories is due to the Scottish philosopher John McTaggart, who first employed it in his paper "The unreality of time", which will be discussed below.

Both theories share an important property: according to each of them, the set of all events[10] is totally ordered by temporal relations. That is, the times of occurrence of any two events can be compared through these relations.

According to A-theories, events are ordered by the predicates present, past, and future. That is, not only is the set of all events made up of three different types of events,[11] but past, present, and future events can themselves be compared with their like using the same relations. For example, any two past events x and y can be compared on the basis of whether one is in the past of the other. Events x and y are in each other's present if and only if neither x is in the past of y, nor y is in the past of x. Also, y is in the future of x if and only if it is neither in the past nor in the present of x. But that is not all: according to A-theory, we are not simply faced with a static set ordered by these quite cumbersome predicates. Rather, the A-predications themselves constantly change: All events which are now in the future will become present and will subsequently flow away into the past. Events in the remote future are constantly flowing into the closer future, and events already in the past are flowing into the more remote past. Thus, A-theory is often termed the "dynamic" view of time.

In the case of B-theories, the total ordering is much simpler: relations such as "before", "after", and "simultaneous with" suffice to establish it. We can simplify this still further and use only the reflexive, antisymmetric

[10] The following discussion will make extensive use of the notion of an "event", and of sets of events, since models of time can be characterized well in these terms. Moreover, these notions are central to relativity theory. Of course, employing this terminology does not commit us to an event ontology whereby events are more basic than things.

[11] Note that, by throwing present, past, and future events into the same set, nothing is prejudiced of their ontological status, which may differ. That is, the set could contain real and unreal members. This is not to say that such a model is unproblematic: if, for example, future events are simply non-existent, in what sense can they be said to approach the present?

and transitive relation *Bxy*, understood to mean "*y* is before or simultaneous with *x*", to compare the times of occurrence of any two events. Then, if *Bxy* and *Byx* are both true, *x* and *y* occur at the same time. The A-theoretical predicates of "present", "past", and "future", and the ontological distinctions associated with them, are entirely dispensed with, so that the structure of time is now analogous to that of an axis of space, whose points can be ordered by their distances from some arbitrarily chosen origin. B-theoretical speech thus takes a "bird's eye" perspective on the course of events.

In addition, according to both theories, it is possible to define a metric on the set of events,[12] in order to measure temporal distances between them. Like with all measurement, an irreducibly conventional element is involved, because the choice of the basic unit of measurement is arbitrary. On the other hand, when ratios between two measured quantities—in this case, temporal distances—are taken, these conventional units cancel and we obtain a pure number. However, on a phenomenological level at least, time behaves A-theoretically, so that temporal measurement is beset by a philosophical difficulty which was acutely perceived by Augustine of Hippo, and which does not occur in spatial measurement: since, for any nonzero temporal distance between two events, at least one of them is in the past or future, it seems that we are measuring things which are nonexistent, either because they no longer exist or because they do not exist yet.[13]

Two other options for the topological structure of time must be mentioned: Conceivably, there might be no mind-independent temporal ordering of events. Alternatively, *partial* temporal orderings may exist on the set of all events, whether those orderings be conceived of as A- or B-theoretical.

This section has provided an outline of the general characteristics of both A- and B-theoretical conceptions of time. However, each theory in turn gives rise to a variety of models of time, which spell out the ontological status assigned to events in time in different ways. This is especially

[12] On defining a metric on a set, cf. F. Hausdorff, *Grundzüge der Mengenlehre* (1978), pp. 211–214.

[13] Augustine concluded that temporal measurement takes place entirely in the mind, exclaiming: "in te ergo, anime meus, tempora metior." *Confessiones*, XI, 27.

true of A-theory, whose proponents face the task of explaining the ontological differences between past, present, and future events.

2.2.2 A-Models

A-models recognize a mind-independent passage of time and an associated change in the existing. While its advocates do not all agree on the ontological status of past and future events, there is consensus among them that present events are real. In addition, the status of the past is different from that of the future on most dynamic accounts.

For example, according to the model of time known as the "growing-block" view, the past and the present are real, but the future is not yet real. As time passes, more and more future events become real, so that successive layers are continually added to the block which represents all that exists. Closely related to this model is that of the "shrinking tree", according to which there is a unique past and present—constituting the "trunk" of the tree—but many, indeed infinitely many, possibilities of what might happen in the future, given the fixed past and present. These alternative courses of events "branch off" the trunk, and this branching is situated at the present moment where, of the myriad possible courses of events, only one is selected. The other branches are, so to speak, cut off and prevented from becoming realized.

Another asymmetric model of time, emphasized and described in detail by Mauro Dorato, must be mentioned in this context: half-open models of the future.[14] According to them, the truth value of propositions about some, but not all, future events are determined at a certain moment which precedes their occurrence. For example, once a person comes into existence, it becomes certain that they will die, although not all propositions about the details of their death have definite truth values at that time. Once the death occurs, it becomes determinate or fully "attribute specific",[15] that is, all relevant propositions about it acquire definite truth values. According to models of a fully open future, an

[14] M. Dorato, *Time and Reality: Spacetime Physics and the Objectivity of Temporal Becoming* (1995), pp. 57–67.

[15] To use the formulation in Dorato (1995), p. 61.

event's becoming real coincides with all propositions about the event acquiring definite truth values, whereas according to half-open models, some propositions about an event can acquire a truth value without the event itself becoming real.

A-models furthermore differ in their answers as to what it means for an event to be past. There is a deeply seated intuition whereby past events are "no longer" real, and must therefore be given a different status from present ones, a distinction which the growing-block and shrinking-tree models by themselves do not afford. Moreover, it would seem that this situation is not adequately modelled by postulating that past events are unreal (like future ones and unlike present ones), but determinate (unlike future ones, but like present ones): for fictitious events too can be viewed as determinate, at least with respect to many propositions, but unreal. Unlike them, however, past events have really occurred, and therefore, it seems, "have been" real although they "are" no longer real. This seems to confer a kind of half-real status to past events, superior to that of the merely possible, the fictitious, and also the future, but inferior to that of the fully real present, leading to the problem of whether a coherent account can be given of this peculiar status.

This difficulty is avoided by the position known as presentism, which asserts that only the present is real, whereas the past and the future are in no way real, or at any rate only as real as merely possible, but not actual states of affairs.[16] An important consequence is that statements about the past are strictly speaking false, since the entities which these statements involve do not exist. That is, there is no mind-independent state of affairs or "truthmaker" in virtue of which such statements could ever be true. According to presentism, it is not true that Napoleon lost the Battle of Waterloo, since Napoleon and the battle do not exist. To avoid this difficulty, many presentists concede that there are presently existing truthmakers for some past events, such as for those which still causally affect the present, or for which there is evidence in the present, so that statements

[16] For a discussion of the meaning of "only the present is real", cf. T. M. Crisp, "Presentism", in M. J. Loux, D. W. Zimmerman (eds.) *The Oxford Handbook of Metaphysics* (2003).

about the past can be grounded.[17] Events which have sunk into "causal oblivion" are deemed on that account to have no truthmakers.

2.2.3 B-Models

The models of time advanced by B-theorists are much simpler, since B-theory is not plagued by the difficult problem of accounting for any differences in the ontological statuses of past, present, and future events. According to B-theory, the time-axis is not in principle different from the three spatial axes, so that the position of an event on the time-axis is of no relevance for its ontological status. In this respect, B-theory is closely related to the classical block universe, according to which there is no change in the existing, but all events are simply given—hence also the alternative name of "eternalism" for this view. In the words of Arthur Eddington, "Events do not happen; they are just there, and we come across them."[18] Because, unlike in A-theory, events at any time are regarded as existing, B-theory in principle allows the possibility of time travel.[19] B-theory is usually also taken to mean that all events in four-dimensional spacetime are fully determinate. For it is widely conceded that, whatever the status of the future, what we perceive as the "past" is at any rate fully determinate, so that what we take to be the "future" is therefore to be assigned the same status.

B-theoretical models are therefore often associated with a view which one could term "temporal determinism". In this context, it is crucially important to distinguish temporal from physical determinism. An event is physically determined exactly if its occurrence is implied by laws and initial conditions, that is, exactly if given those laws and initial conditions, the event in question must occur. To define temporal determinism, it will be necessary first to give a precise definition of the term "determinate", which was left undefined in the preceding discussion: an event

[17] Dainton (2010), pp. 90–91.

[18] Eddington's block-universe view will be discussed in more detail in Sect. 4.2.

[19] This is not to say that all B-theorists hold time-travel to be possible: see, for example, D. H. Mellor, *Real Time II* (1998), pp. 125–135. On A-theorist views of time travel, see Dainton (2010), pp. 138–141.

is henceforth taken to be determinate with respect to a proposition describing it if that proposition has a definite truth value. That is, such a proposition is either true or false. As a counterexample, consider the proposition "there will be a sea battle tomorrow". Aristotle held that this proposition does not have a definite truth value—the sea battle is indeterminate with respect to it.[20] An event will be called "fully determinate" if *all* propositions which can meaningfully be used to describe it have a definite truth value.[21] Alternatively, we might also say that all questions which can meaningfully be asked about a fully determinate event have a definite answer.

We can now define temporal determinism: it holds true of any event e with respect to some proposition p at a time t exactly if e is determinate with respect to p at time t. This also implies that e is determinate with respect to p for all times later than t. For if the truth value of p were to change after t, both p and not-p would be true of e. For example, the proposition "the Greeks will win the sea battle in AD 2100", if true today, cannot be false tomorrow. Hence, if temporal determinism is true of an event with respect to some proposition, then the event is fixed and unchangeable with respect to it.

If physical determinism holds for e with respect to p at t, then temporal determinism holds as well, since the truth value of p is determined by physical laws and initial conditions at t. But the converse is not true: temporal determinism does not imply physical determinism, since the definite truth value of p describing e at t may not be due to laws and initial conditions. Karl Popper illustrates this point with the example of a film: When we view the opening scene, the relevant truth values of propositions describing the next scene are already definite. However, unlike in physically deterministic situations, the content of the next scene cannot be predicted from that of the opening scene by using a set of laws.[22]

[20] *De Interpretatione*, 19a,1–19b,4.

[21] The addition "which can meaningfully be used to describe it" is necessary, since not all predications can meaningfully be made of all events.

[22] K. Popper, *The Open Universe: An Argument for Indeterminism* (1988), pp. 5–8.

Note that the definition of "temporal determinism" only serves the purpose of illustrating a certain model of time. I will argue later that there is no temporal determinism over and above causation. That is, truth values of propositions about events are not determined in virtue of their position in time.

In the classical block universe as described above, *temporal* determinism holds for all events with respect to all relevant propositions describing them, at any time, whether or not this is due to *physical* determinism. B-theory, on the other hand, requires only that the ontological status of events not depend on their position on the time axis, but not necessarily that all events must be regarded as determinate in all respects. This opens up another option: if it is possible for events to be indeterminate with respect to a proposition which can meaningfully describe such events,[23] then B-theory would also allow such "fuzzy" or "blurred" events to be spread throughout spacetime, including the regions which, based on our everyday intuition, we would qualify as "past". I will term this model the "fuzzy block".[24] Since, however, it is generally conceded that some events are determinate with respect to at least many (or perhaps all) relevant propositions describing them, it might be more accurate to speak instead of the "half-fuzzy block".

For completeness, a further important feature of B-theoretical models must be mentioned: since the time axis in a B-world is not in principle different from any spatial axis, it would follow that temporal parts of an object (the stages or phases of its existence, defined by their initial and final points in time), must be regarded in the same way as its spatial parts. This leads to the view of physical objects as four-dimensional "spacetime worms" embedded in the surrounding spacetime.[25]

[23] Such a possibility may well not have been taken seriously prior to the advent of quantum mechanics. Whether this theory allows it is open since, for example, it can be argued that it is simply not meaningful to assign a particle both a clearly defined value of position and momentum—or of any other pair of canonically conjugate variables—at any particular time.

[24] Note that the "fuzziness" refers to the absence of definite truth values, not to physical indeterminism, although the latter is arguably necessary for there to be any such fuzziness.

[25] Runggaldier, Kanzian (1998), pp. 100–104.

2.2.4 A Brief Assessment of A- and B-Models

The strengths and weaknesses of A- and B-models can now be assessed. The former, generally speaking, capture our everyday understanding of the passage of time and the asymmetry between past, present, and future well. But they are faced with massive problems in spelling out what the A-predicates "past", "present", and "future" precisely are supposed to mean. This problem, as we have seen, is especially pressing concerning the ontological status of the past. The solution offered by presentism pays the price of failing to distinguish the ontological status of the past from that of the merely possible, or indeed the fictitious. The growing-block model recognizes an equal status of existence for the past and the present. It thus avoids the problem of having to acknowledge a half-existing realm, thereby sacrificing the distinction between the ontological statuses of the present and the past.[26] By and large, A-theory seems to leave us with a mess of quandaries.

But the most important challenge to A-theoretical conceptions of time is the well-known argument devised by John McTaggart himself in his article "The unreality of time"[27]: in the first step of this argument, McTaggart charges that, according to A-theory—where all events flow from the future through the present and into the past—one and the same event is assigned three mutually incompatible predicates, namely those of being future, present, and past. The response of the A-theorist is not hard to guess: the three predicates are indeed incompatible, but they are not assigned to the same event in the same respect, so that there is no contradiction. For example, an event which is now in the past once was present, and before the time of its occurrence was a future event. In short, incompatible predicates are assigned to the same event, but at different times, so that it is wrong to say that the event in question is past, present, and future *simpliciter*. However, at this point McTaggart counters that the

[26] Although one might object that the present is still distinguished from the past in a very relevant sense according to the growing-block model: it is the last layer of the block, set between the fixed and real past, and the open and unreal future, thus making it, so to speak, the locus of becoming.

[27] This article was first published in *Mind: A Quarterly Review of Psychology and Philosophy* (1908) 17, 457–474. A revised version thereof, drawn from McTaggart's *The Nature of Existence*, can be found in R. Le Poidevin, M. MacBeath (eds.) *The Philosophy of Time* (1993).

contradiction is not solved, but merely shifted: according to the A-theorist, the time of occurrence of the event is past in the present, present in a particular past, and future in a still more remote past. Thus, it is now the time of occurrence which is given the incompatible predicates past, present, and future. Again, the A-theorist could reply that the time of occurrence is not assigned these three predicates simultaneously, but at different times. But this only shifts the problem to the time of the time of the occurrence, and so forth. Thus, A-theory can only avoid contradiction by incurring an infinite regress. The conclusion is that A-theory, while intuitively plausible, is in fact inconsistent and must be rejected.

McTaggart goes still further: according to him, B-relations such as "before" themselves depend upon A-relations. Thus, if A-theory is false, there is nothing to ground B-relations, and therefore there is no temporal ordering of events. However, according to McTaggart, all events are nevertheless ordered, although this ordering is not of temporal nature. Such an ordering, called a C-series, has no intrinsic direction, unlike the B-series, which is marked by a forward direction in time.

A well-known response to McTaggart's argument is that by C. D. Broad:

> I cannot myself see that there is any contradiction to be avoided. When it is said that pastness, presentness, and futurity are incompatible predicates, this is true only in the sense that no one term could have two of them *simultaneously* or *timelessly*. Now no term ever appears to have any of them timelessly, and no term ever appears to have any two of them simultaneously. What appears to be the case is that certain terms have them *successively*. Thus there is nothing in the temporal appearances to suggest that there is a contradiction to be avoided.[28]

Broad's objection to McTaggart's argument captures the commonsense view of events successively changing from being future to being present, and then to being past, a view which, as it at first sight seems, is misrepresented by McTaggart's contention that events have three incompatible properties *simpliciter*. However, as McTaggart himself pointed out, this move only shifts the problem of incompatible properties to the

[28] "Examination of McTaggart's Philosophy", in Van Inwagen, P., Zimmerman, D. W. (eds.) *Metaphysics: The Big Questions* (1998), here p. 77. Emphases in the original.

times at which events occur. Broad's response does not offer any escape from the regress incurred in this way.

A way out of McTaggart's paradox is offered by presentism: since according to it, everything which exists is present, no event is ever past or future, so that nothing exists which has incompatible temporal predicates.[29] This does not, however, make presentism a viable option. For on it, past entities, whether events or things, are simply unreal. Caesar then has the same ontological status as a character in a novel. Even versions of presentism which allow truthmakers for past entities, so that statements about them can in some sense be grounded, hold these entities themselves to be unreal. Given presentism, we can give no account of the evolution of objects, so that reality becomes an unintelligible succession of "nows" with nothing to ground a causal connection between them. Furthermore, as will be argued throughout this book, the notion of a global present or "now" cannot be defined in an observer-independent way, but rather relies on an extrapolation of a feature of observers' experience.

Note that we can also avoid McTaggart's paradox by embracing a growing-block model, according to which the future is unreal, whereas present and past events have the same ontological status. Then, there are again no events with incompatible properties—although of course an account would still need to be given as to why the past is gone and unavailable, or at least seems to be so. However, the growing block is not the option I will advocate, because no observer-independent and global distinction between its two regions, the real past-present, and the unreal future, is available.

B-theory nicely circumvents the problem of the different ontological status of past, present, and future events. However, since it puts the time axis on a par with the spatial axes, it fails to capture features of time which are difficult to deny: the unidirectionality of its flow, which ensures that no point in time can be revisited twice, and the overwhelming impression of a significant difference between present, past, and future. Furthermore, there is the difficulty already mentioned concerning the ontological

[29] Cf. the discussion in M. Moling, *Zeit und Ewigkeit nach Thomas von Aquin*, PhD thesis at Innsbruck University (2005), pp. 32–38.

foundation of B-relations: if time does not flow, what grounds relations such as "before"?

A- and B-theoretical models of time can also be assessed on theological criteria: the former, with their characteristic asymmetry between the fixed past and the open future, seem to be more hospitable to free agents responsible for their own actions. It is perhaps for this reason that they are supported by many contemporary Christian philosophers, as we will see. On the other hand, the continuously changing "now" in such models leads to problems such as, for example, the dilemma raised by Norman Kretzmann: how can God know what time it is now without being subject to change? But if God is subject to change, how can He be a perfect being?[30] Conversely, B-theoretical models accommodate no ontologically robust asymmetry between past and future required, as it seems, for free agents. On the other hand, Kretzmann's problem doesn't appear given them, since there simply is no "now" for God to know. In Chaps. 7, 8, 9, 10 and 11, I will argue in favour of a model beyond A- and B-theory, in which temporal action asymmetry requires no such absolute "now". This result will simplify problems on the relationship between God and time, to which we will return in Chaps. 12, 13 and 14.

2.2.5 The Instant Moment

As already emphasized, according to A- and B-theories, at least in their original formulation, there is a total ordering on the set of all events. Thus, according to both theories, for any time of occurrence of an event, there is a uniquely defined set of all events happening at that time. If spacetime is represented in a four-dimensional diagram, with three spatial axes and one temporal one, such a set constitutes a three-dimensional slice of spacetime, or more precisely, a hyperplane[31] embedded in it, whose equation is simply $t = constant$. This hyperplane is a "snapshot" of the universe, representing its state at a given instant. For A-theory, this means in particular that there is a uniquely defined set of all events

[30] "Omniscience and immutability", *The Journal of Philosophy* (1966), here p. 409.

[31] A hyperplane in an n-dimensional vector space is a vector sub-space of dimension $n-1$ embedded therein—just like a two-dimensional plane is embedded in three-dimensional space.

occurring *now*. Moreover, in both A- and B-theory, it follows that there is a total ordering of such hyperplanes, representing the history of the entire cosmos.

In this context, we are assuming that events are point-like in time, that is, of infinitesimal duration. As for temporally extended events, which are then better called "processes", given both A- and B-theory it is always possible to identify the parts thereof which take place at the time in question.

This picture matches the everyday understanding of time, where it is generally assumed that the notion of "the set of all events which occur at a given moment" is unproblematic. It is also a consequence of Newtonian physics, since Newton, as already mentioned, assumed an "equable" flow of time throughout the universe, which does not depend on the content of the universe.

According to A-theory, the worldwide "now" is the locus where the conversion of possible world states into actual ones occurs, and can thus aptly be called the "hyperplane of becoming". This view is well expressed by the philosopher John Lucas:

> Quantum mechanics, if it is to be interpreted realistically, distinguishes a probabilistic future of superimposed eigen-states from a definite past in which each dynamical variable is in one definite eigen-state, with the present being the moment at which—to change the metaphor—the indeterminate ripple of multitudinous wave-functions collapses into a single definite wave. Admittedly, many of those who think about quantum mechanics are not realists, and admittedly again, there are horrendous difficulties in the way of giving a coherent account of the collapse of the wave-function. But an obstinate realism, as well as a slight sympathy for our feline friends, precludes my envisaging any long period in which Schrödinger's cat could be half-dead and half alive, and this whether she be in a laboratory in Europe or on some planet circling Betelgeuse. There is a definite fact of the matter, there as much as here, whether or not we are dealing with a superposition of functions or one definite eigen-function. And hence there is a unique hyperplane advancing throughout the whole universe of collapse into eigen-ness.[32]

[32] "Transcendental Tense" (1998). http://users.ox.ac.uk/~jrlucas/time/transten.html.

It is often assumed that such a hyperplane of becoming, which separates past from future regions of spacetime in an unambiguous way, is a necessary condition for there to be a truly dynamic, evolving universe and that the lack thereof would result in a static block universe. I will argue that this is not so.

How are the slices of cosmic history related to each other, and what is it that brings about the next slice, given the previous one? This question is simply a version of the problem of change, but this time applied to the universe as a whole. If the ordered hyperplanes together constitute the history of the universe, we would expect there to be a relationship of both continuity and discontinuity between the slices.

An illustrative simple case of such a relationship is the one which obtains in a deterministic universe as conceived by Pierre Simon de Laplace. Here, any future state of the universe is determined entirely by any past state together with the laws of nature. In other words, the laws of nature together constitute a function which takes as input a world state and produces the next world state, from which in turn the next world state follows by re-application of the same function, and so on for all subsequent world states. Operationally, this means that, given a sufficiently detailed kinematic description of the universe at an instant of time,[33] it is in principle possible to predict any future state of the universe to any desired degree of precision. Laplace assigned this task to a hypothetical being of vast intelligence and seems to have suffered no illusions that such a feat would ever come within the reach of human technology.[34]

---

[33] The concept of a "kinematic description of the universe at an instant" suffers from the difficulty that instantaneous velocities need to be assigned to particles, but such velocities themselves are only obtained through a limiting process, so that in practice a time-slice of finite width will need to be taken into account. Cf. the discussion in Dainton (2010), pp. 290–294.

[34] "We must therefore regard the present state of the universe as the effect of its precedent state, and as the cause of the one which will follow. An intelligence which, for a given instant, would know all the forces by which nature is moved, and the respective positions of the beings which compose it, if furthermore it were vast enough to submit these data to analysis, would comprehend within the same formula the movements of the greatest bodies of the universe, and those of the lightest atom: nothing would be uncertain for it, and the future as well as the past would be present to its eyes. The human spirit, in the perfection which it has been able to give to astronomy, presents a feeble shadow of this intelligence … All its efforts in the search for truth tend to bring it ever closer to the intelligence which we have just devised, but it will always remain infinitely distant from it." In the original: "Nous devons donc envisager l'état présent de l'univers, comme l'effet de son état antérieur, et comme la cause de celui qui va suivre. Une intelligence qui pour un instant donné,

As Karl Popper points out, such a being does not need to be omniscient. It is sufficient that it be able to carry out predictions in the way just described.[35] In short, while a Laplacian universe evolves in time, the truth values about all future events are already fixed throughout the history of the universe.[36]

connaîtrait toutes les forces dont la nature est animée, et la situation respective des êtres qui la composent, si d'ailleurs elle était assez vaste pour soumettre ces données à l'analyse, embrasserait dans la même formule, les mouvemens des plus grands corps de l'univers et ceux du plus léger atome: rien ne serait incertain pour elle, et l'avenir comme le passé, serait présent à ses yeux. L'esprit humain offre dans la perfection qu'il a su donner à l'astronomie, une faible esquisse de cette intelligence… Tous ses efforts dans la recherche de la vérité, tendent à le rapprocher sans cesse de l'intelligence que nous venons de concevoir, mais dont il restera toujours infiniment éloigné." P. S. Marquis de Laplace, *Théorie Analytique des Probabilités* (1814), pp. ii–iii.

[35] Popper (1988), p. 30.

[36] There are, however, difficulties with the notion of a deterministic universe evolving in time at all, as will be discussed in Sect. 6.2.

3

The Impact of the Special Theory of Relativity on the Philosophy of Time

The special theory of relativity (henceforth SR) presents massive challenges to an "everyday" understanding of time, but it also furnishes some invaluable insights. Much has been written on the subject of just what conclusions philosophy of time ought to draw from SR, and which *non sequiturs* must be avoided. This chapter offers an overview of the main points of contention in this area, and an assessment of the key arguments advanced. Appendix A contains an outline of SR for readers not acquainted with this theory.

3.1 Immediate Consequences of SR for the Philosophy of Time

3.1.1 Loss of the Worldwide Instant

The Lorentz transformations are equations in SR which allow switching from one inertial frame of reference to another.[1] From them, it emerges

[1] Given at the end of Appendix A. I am adopting the convention of setting $c = 1$.

D. Saudek, *Change, the Arrow of Time, and Divine Eternity in Light of Relativity Theory*, Palgrave Frontiers in Philosophy of Religion,
https://doi.org/10.1007/978-3-030-38411-1_3

that if $t' = 0$, $t = \beta x$. In words, if an observer (in this equation, the primed one) passes another (the unprimed one) at a fraction β of the speed of light, the former's axis of simultaneity has slope β relative to that of the latter. Likewise, the time axis of the primed observer is given by $x' = 0$, so its equation is $t = \frac{1}{\beta}x$ in unprimed coordinates. However, the boundaries of a light cone are independent of the speed of the frame of reference due to the fixity of c. For two observers passing each other at relative speed $\beta = 0.25$, we therefore get a picture as illustrated overleaf. Thus, there is no unique hyperplane defining the set of all events occurring in the universe at a given moment. Rather, the hyperplane of simultaneity for an observer is determined by his state of motion. In fact, since an observer is allowed to travel at any speed $\beta < 1$, that is, any speed strictly less than that of light, the axis of simultaneity can have any slope less than one. Therefore, it seems that B-relations such as "before" do not establish a total ordering of events and that there is no observer-independent temporal order of world states. In particular, the loss of a uniquely defined hyperplane of simultaneity massively challenges the notion of a worldwide "now", and thereby also that of an advancing hyperplane of becoming which would separate the determinate past from the open future (Fig. 3.1).

In short, both A- and B-theory, either of which assumes a total temporal ordering of the set of all events, are in trouble. Defenders of an observer-independent simultaneity, or even of a worldwide "now", could of course still claim that there is an ontological foundation for these notions elsewhere but, at any rate, no support seems to be forthcoming from spacetime physics, as will be examined in more detail below.

What is more, it seems that an account can be given of how human beings have developed the so deeply seated intuition of a uniquely defined present: due to the vast speed of light compared with velocities of objects we experience in everyday life, and due to the relatively short distances of our everyday surroundings, we are immediately informed by light signals about what is going on around us, thus giving rise to the notion of "all that is happening now". The fact that we are so small compared to cosmic scales thus allows us, through the sense of vision, to react to our environment with hardly any time delay. As R. J. Russell puts it: "Clearly the

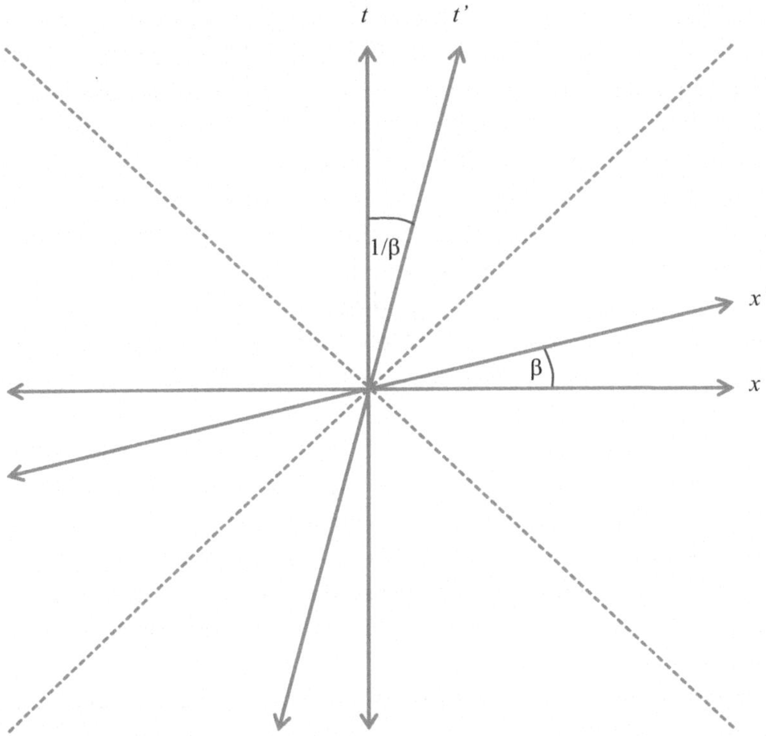

Fig. 3.1 A spatial axis and the time axis in two inertial frames of reference

so-called global present of our everyday experience is an anthropocentric and mythological illusion that arises from the relatively slow velocities of our ordinary experience compared with *c*."[2]

3.1.2 Invariant *K*-relation

According to SR, two events are causally connectable exactly if light has enough time, or more than enough time, to travel between both. Mathematically, the spacetime interval between them is then greater than or equal to zero. Causal interactions in spacetime furthermore exhibit a

[2] R. J. Russell, *Time in Eternity: Pannenberg, Physics, and Eschatology in Creative Mutual Interaction* (2012), pp. 264–265.

very familiar, yet peculiar, asymmetry: earlier events can affect later ones, but not vice versa. For example, we-now can influence events which are timelike-separated from us and in the future, by sending some signal to the location of these events, but the reverse is not true. For now, let's take this asymmetry for granted. It is then possible to define a relation *Kxy*, meaning "*x* can be causally affected by *y*", which holds between some events in spacetime. A necessary condition for it to obtain is that the spacetime interval between *x* and *y* be greater than or equal to zero. But this is not sufficient: it is also required that *x* occur later than *y* according to some clock. Then, *x* will occur later than *y* according to all frames of reference.[3] In other words, *Kxy* holds exactly if *x* is within or on the future light cone of *y*.

The relation *K* is irreflexive, that is, no event can causally influence itself. It is asymmetric: if *x* can be causally affected by *y*, then *y* cannot be causally affected by *x*. Moreover, it is transitive: if *y* can be affected by *x*, and *z* can be affected by *y*, then *z* can also be affected by *x*. This can be seen by drawing *y*'s future light cone: since it lies entirely within the future light cone of *x*, any event *z* in *y*'s absolute future is also in *x*'s absolute future.

Note that *K* has the same relational properties as the temporal relation "later than", and that at least the properties of irreflexivity and of asymmetry are shared by the relation "being causally dependent on", although the transitivity of this latter relation is debatable.[4]

The *K*-relation thus fixes the direction of the time axis in a spacetime diagram. This direction, unlike that of the spatial axes, is not arbitrary, but can be thought of as that of "forward causation". If, as is widely held, causes precede their effects, and if *x* is, or can be, causally affected by *y*, then *x* must be later than *y*, and must therefore have a greater time coordinate.

---

[3] The reasons are given in Appendix A, and the philosophical difficulties associated with switching the temporal order of causally connectable events will be discussed in more detail in Sect. 11.2. Note also that *x* and *y*, here referred to as "events", can of course simply be thought of as points situated in spacetime, whether or not there is anything noteworthy occurring there. We might therefore alternatively say that *Kxy* holds between any two points *x* and *y* in spacetime exactly if whatever is going on at *y* can influence what happens at *x*.

[4] Cf. M. Tooley, *Time, Tense and Causation* (1997), p. 255.

Difficult issues concerning the *K*-relation remain, however. The relation of being causally connectable between events—that is, of being timelike or lightlike-related—is symmetric, and furthermore, the temporal order of events makes no difference as far as this relation is concerned, since replacing the time interval *t* between two events by –*t* does not affect the spacetime interval. Whence then the familiar asymmetry whereby *Kxy* implies not-*Kyx*? This problem has been explored by John A. Wheeler and Richard P. Feynman, who explained it in terms of an underlying time-symmetric interaction, that is, one where electromagnetic waves travel both forward and backward in time from the emitter, but in such a way that the backward-travelling wave cancels.[5] We will revisit the *K*-relation in Sect. 11.2, where I will propose a way of reconstructing this asymmetric relation, *without* however assuming that causes precede their effects.

However, for the purpose of this section, it suffices to note that the *K*-relation ensures an invariant relation of temporal asymmetry between some pairs of events, that is, one which is independent of the frame of reference. Moreover, if relativity theory holds in the entire universe, and if, as seems plausible, each event *x* is connected by the *K*-relation to some other event *y* (i.e. either *Kxy* or *Kyx*), then each event *x* is a member of *some* set of events ordered by it, and ordering which we might call a "*K*-chain". But in general, *K*-chains cannot themselves be ordered in this way. For example, if *x* and *y* are spacelike-separated, and if we choose a *K*-chain starting at *x* and extending into its causal future, and another such chain starting at *y*, then these two chains do not stand in a *K*-relation to each other. Thus, the overall picture seems to be that all events are members of some partial temporal ordering on the set of all events brought about by the *K*-relation, not of a total temporal ordering on the set of all events in spacetime.

To sum up, SR furnishes arguments both for the proponents of a changeless view of the universe as well as those of a dynamic view: The former can point to the impossibility of defining what notions such as the "present", "past", or "future" of the universe are supposed to mean

[5] "Interaction with the absorber as the mechanism of radiation", *Reviews of Modern Physics* (1945). For a brief, nontechnical exposition, see R. J. Russell (2012), pp. 340–347.

independently of the state of motion of an observer. The latter can point to the *K*-relation holding between some pairs of events independently of the frame of reference, and can argue that, given it, temporal asymmetry is an objective feature of the world.

3.2 Arguing from SR to Temporal Determinism

It has frequently been argued that the loss of the worldwide "now" leads to a view where by all events in spacetime have the same determinate status, so that the universe is a changeless block. To assess it, it will be useful to consider the arguments put forward in its favour by Wim Rietdijk and Hilary Putnam,[6] as well as the objections against these arguments, especially those advanced by Howard Stein.[7]

Rietdijk and Putnam provide, independently of each other, succinct and simple arguments which, while differing in a few details, can both be summed up as follows.

For any pair of events x and y, if y is simultaneous with x, or in the past of x, then it is fully determinate as of x, i.e. all relevant truth values about it are already definite. Then, for an observer for whom x happens now or is past, y is determinate. Moreover, if an event is fully determinate for one observer, it is fully determinate *simpliciter*, since "being determinate" is an observer-independent property of an event. For example, if it is determinate as of x that the Greeks are winning a certain sea battle, then the truth value of the proposition "the Greeks are winning the sea battle" is definite, and cannot therefore be indefinite for an observer for whom the sea battle is in the future of x, regardless of whether or not they know the outcome of the sea battle. It is concluded that "being determinate as of" is a transitive relation, i.e. if z is determinate as of y, and y is determinate as of x, then z is determinate as of x.

---

[6] C. W. Rietdijk, "A rigorous proof of determinism derived from the special theory of relativity", *Philosophy of Science* (1966). H. Putnam, "Time and physical geometry", *The Journal of Philosophy* (1967).

[7] "On Einstein-Minkowski space-time", *The Journal of Philosophy* (1968); "On relativity theory and openness of the future", *Philosophy of Science* (1991).

Next, let *a* be an event situated at the origin of some inertial frame of reference *A*, and which is considered a present event in that frame. Let *c* be an event in the absolute future of *a*, and *b* one situated in the elsewhen of both *a* and *c*, as illustrated below. Then, there is a frame of reference for which *a* and *b* occur simultaneously, and another one for which *b* and *c* occur simultaneously. Therefore, *b* is determinate as of *a*, and *c* is determinate as of *b*, and by transitivity *c* is also determinate as of *a* (Fig. 3.2).

Note that, by choosing appropriate inertial frames, it is possible to build a "ladder of transitivity", so to speak, to any event *c* in the future light cone of *a*, no matter how far away in the future *c* might be. For we can choose an event *b* arbitrarily far away in spacetime from *a*, and we can also, by choosing the right velocity, make our plane of simultaneity approach arbitrarily closely to the boundary of *a*'s future light cone, so that there is no limit to the positive time coordinate—according to frame of reference *A*—which the event *b* might have. In Dorato's words, "any future-pointing timelike vector can be written as the sum of two spacelike

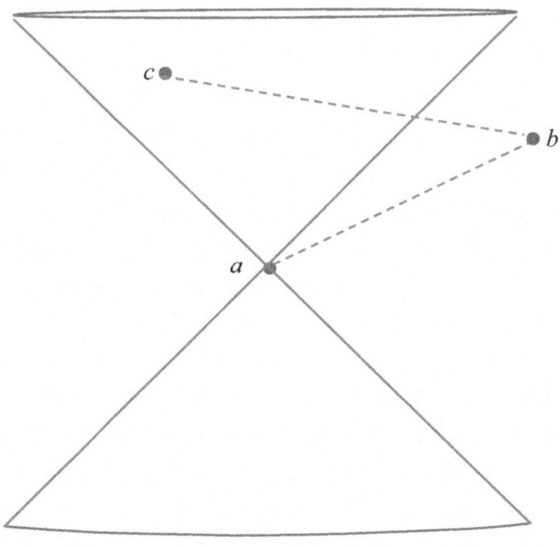

Fig. 3.2 The transitivity of "being determinate as of"

vectors."[8] Of course, we could also build our ladder by using several intermediate events, in a variety of different ways.

In Rietdijk's version of the argument, event *b* is spatially removed from *a*, and simultaneous with it according to frame of reference *A*, whereas *c* is simultaneous with *b* according to some other frame of reference in uniform motion relative to *A*. This, given the above premises, establishes that *c* is determinate as of *a*. Rietdijk concludes that an observer at spacetime point *a* "could not possibly have influenced" events in his own future light cone "in an arbitrary way",[9] since what is going on there can legitimately be considered part of the fixed past by some observer at some spacetime point which is present as of *a*. Rietdijk then gives additional emphasis to this argument by introducing an observer who suddenly changes his state of motion: this observer is first in uniform motion relative to *A*, in such a way that a certain event in event *a*'s absolute future, which we have called *c*, is present for him. Such an observer could suddenly decelerate in such a way that event *a* now becomes a present event for him. But then, both *a* and *c* are present for such an observer, at almost the same moment.

The details of Putnam's argument are slightly different from those of Rietdijk's: Putnam starts with the observer "I-now" at the origin of a coordinate system: "I-now am real." He then introduces a relation *Rxy*, understood as "the relation of simultaneity-in-the-observer's-coordinate-system". Next, another observer "you" is considered, such that "you-now and I-now are at the same place now, but moving with relative velocities". Suppose I-now thought that "all and only those things which stand in *this* relation *R* to me-now are real". Given the principle that there are no privileged observers, not only must you-now be acknowledged as real, but also everything standing in the *R*-relation to you-now—and this includes events which are not *directly* *R*-related to me-now. Thus, "since the event *X* bears the relation *R* to you-now, and you-now bear the relation *R* to me-now, the event *X* bears the relation *R* to me-now." Some such events will be in the future of me-now, that is, their time coordinate according to I-now's frame of reference is positive. Thus, I-now's initial assumption must be false. Of course, such events are still located in the

[8] *Time and Reality* (1995), p. 167.

[9] Rietdijk (1966), p. 342.

elsewhen, rather than the *absolute* future of me-now. But this does not save the absolute future from determinism: for the principle that there are no privileged observers "requires that I also count every thing and event which bears the *transitive closure* of *R* to me ... But every thing and event in space-time bears the closure of the above relation *R* to me."[10]

In short, if the argument advanced by Rietdijk and Putnam is correct, the entire future of every event is determinate, and we are living in a static block universe. Thus, according to Putnam "Aristotle was wrong" because "contingent statements about the future already have a truth value ... the problem of the reality and the determinateness of future events is now solved".[11] Rietdijk is no less clear: "Thus each event is determined: it is already 'past' for someone in our 'now'." Hence, "there is determinism ... also in microphysics. Also there is no free will ... existentialism is untenable."[12]

It is crucially important to note that the determinism argued for by Rietdijk and Putnam is of the temporal, not the physical type, since it is based on the already determinate truth values of propositions about future events, not on initial conditions together with deterministic laws. If their argument is correct, determinism holds also in a universe in which Laplacian determinism is not true, provided of course that SR holds.[13]

3.3 Objections Against Relativistic Temporal Determinism

Not surprisingly, the above arguments have attracted criticism. Especially noteworthy is the detailed critique undertaken by the philosopher of science Howard Stein. Let us consider some of the principal objections:

[10] Putnam (1967), pp. 240–243 and 246. Emphases in the original.

[11] pp. 244–245.

[12] Rietdijk (1966), pp. 341 and 343.

[13] As a side note, K. Popper has argued that SR in fact rules out the possibility of a Laplacian demon, and hence of a physically deterministic universe, since the information necessary in order to predict an event will only be available for an observer in the event's future. "Thus the demon of special relativity is no longer that of Laplace; for this demon, as opposed to Laplace's, cannot predict; he can only retrodict." See K. Popper, *The Open Universe: An Argument for Indeterminism* (1988), pp. 57–62.

First, it is doubtful whether the relativistic concept of simultaneity can be equated with our everyday concept of being "present". Stein draws attention to the original meaning of the Latin "prae-sens": what is present is "at hand" or "in front of" one.[14] This cannot be said of simultaneous events in general. Thus, if present events are fully determinate, can we infer that all simultaneous events are so too? Rietdijk's paper somewhat misleadingly employs the notion of an observer's "experiencing the same present", as though the plane of simultaneity could be likened to the experienced present—although Rietdijk is of course fully aware that the question which events are to be counted as present with which can only be answered after the fact, once all relevant light signals have reached the observer.[15] More generally, Rietdijk and Putnam carry over pre-relativistic temporal concepts, that is, those of the determinate past and present, into the context of Minkowski spacetime, in order to reach their conclusion. In Stein's words, "The common fallacy of those arguments is their employment, in the context of Einstein-Minkowski theory, of notions of time that are illegitimate in that theory."[16]

It can of course be countered that, if simultaneous events are not granted the same status as present ones, then with what right? By what criterion can the determinate events in our elsewhen—which, presumably, exist—be set apart from the postulated indeterminate ones? Proposals for answering this question will be discussed in the next section. Moreover, if critics of the Rietdijk-Putnam argument wish to save an ontological asymmetry between past and future, their objection of illegitimate carrying-over cuts both ways: are they willing to concede that the present and the past may in fact not be determinate after all? If so, the asymmetry between determinate past and open future is again sacrificed, and it seems that we are back in the block universe, where all events have equal status—although the block would this time be of the fuzzy variety. Indeed, Stein considers the possibility of indeterminateness of events extending also into the past: "inference [from the present to the past], when it is founded on the laws of quantum mechanics, is subject to the

[14] Stein (1991), p. 159.

[15] Rietdijk (1966), p. 342.

[16] Stein (1968), pp. 15–16.

same limitation to *statistical* assertions as is quantum-mechanical inference to the future … [this] point suggests the notion of a progressively increasing *physical uncertainty* (or 'indeterminism') of the past: a kind of fading of the world's memory."[17]

Furthermore, Rietdijk and Putnam argue in favour of universal determinism on the basis of temporal concepts. But "determinism" is, arguably, a physical and causal notion, not a temporal one, since the truth values of propositions about events depend on the relevant causal interactions. Stein puts forward a similar objection: "I do not think the notion of truth as a 'timeless' predicate says anything about the world … determinism apart from causality is a pseudoproblem."[18] As pointed out above, for Rietdijk an observer "could not possibly have influenced … in an arbitrary way" events in his own future light cone. This seems to fly in the face of the theory we are assuming to start with, for according to SR, the future light cone of an event is of course precisely the region which *can* be influenced by that event. The argument whereby events are unalterable because they are already simultaneous to an observer who can be considered present can therefore be likened to the much more ancient fallacy of concluding from God's universal knowledge of all events—including future ones—that all events are unalterable.[19] In logic, the fallacy of concluding from the truth of a proposition to its being true necessarily is known as the conclusion from the *necessitas consequentis* to the *necessitas consequentiae:* for a statement S about a proposition (e.g. let S be "Socrates is sitting"), we have $\Box(S \rightarrow S)$, where $\Box$ is the modal operator of necessity. That is, it is necessarily true that, if Socrates is sitting, he is sitting. But the fact that Socrates is sitting does not imply that he is sitting necessarily. In symbols, it is not true that $S \rightarrow \Box S$.[20]

Finally, the Rietdijk-Putnam argument is one from parity of reasoning[21] and therefore does not rigorously prove what it claims to: It rests on the claim that events in the elsewhen should be acknowledged

---

[17] Stein (1968), pp. 22–23, emphases in the original.

[18] Stein (1968), pp. 21–22.

[19] Discussed by Augustine of Hippo in his *De Civitate Dei*, V, 9.

[20] W. Löffler, *Einführung in die Logik* (2008), pp. 216–217.

[21] Cf. Dorato (1995), pp. 69–70.

equal ontological status to the here-now. But it does not follow that all events in the elsewhen do in fact have this equal status. Perhaps, one might conjecture, there is indeed a boundary in spacetime between the determinate and the indeterminate, where the transitivity of "being determinate as of" breaks down—the problem being only that the Minkowski diagram is blind to such a boundary. Again, if the critic wishes to save the idea that the future can be influenced from the here-now, such a boundary is subject to an important constraint: it must pass through the here-now.[22] For otherwise, observers find themselves either in the fixed past or the open future, and their point in spacetime is not part of the privileged hypersurface[23] where true becoming occurs.

3.4 Carving up Spacetime

Can spacetime be divided into regions of different ontological status in any plausible way, in order to save the asymmetry between past and future? Dorato fittingly likens this to moving "old furniture", that of our pre-relativistic temporal categories, into a "new place", namely relativistic spacetime.[24] He discusses several proposals for doing so: One radical option is the *hic et nunc* view. It is "motivated by strict verificationism" in the same way as pre-relativistic presentism, which regards the past and the future as unreal. In a relativistic context, the strict verificationist must "by parity of reasoning" assign the same status of unreality to those events in their present which are inaccessible to them. The result is a view whereby "only the infinitesimal region of spacetime around a point is real".[25] Dorato notes that, if such a view is combined with the "democratic" principle whereby "every spacetime point is a legitimate here-now", we get just the opposite picture: every event in spacetime can now

[22] Cf. Stein (1991), p. 148.

[23] I will use the term "hypersurface" rather than "hyperplane" for the more general case of a not necessarily flat (n-1)-dimensional surface embedded in an n-dimensional space.

[24] Dorato (1995), p. 131.

[25] pp. 137–139.

with equal right be viewed as real. Thus, there is a "characteristic oscillation between the *hic et nunc* view and the full view" of spacetime.[26]

The spacetime solipsism of the *hic et nunc* view, like any solipsism, cannot be refuted, but can nevertheless be rejected out of hand, as it prohibits any exploration of the existing beyond the immediately observable. Dorato suggests four more options for localizing the determinate and the indeterminate in spacetime, relative to an event *a*, where it is assumed that events in *a*'s past light cone are determinate, and that *a* lies on the hypothetical boundary between determinate and indeterminate:[27]

1. "The first option is to regard as determinate all events below the simultaneity line with respect to an arbitrarily selected inertial frame centered in *a*."
2. All events in the elsewhen of *a* are determinate.
3. All events in the elsewhen of *a* are indeterminate.
4. The answer to the question is underdetermined by data.

All of these options incur grave difficulties: According to option 1, the division between the determinate and the indeterminate arbitrarily depends on the state of motion of a particular observer. Only the hyperplane through *a* associated with that state of motion can then be the true boundary between the determinate and the indeterminate. For if we attributed this privileged status also to a different hyperplane of simultaneity through *a*, associated with a different state of motion, this would lead to a contradiction, since some events which are determinate for one observer would be indeterminate for the other. Since "being determinate" is an intrinsic property of an event, this is impossible. The only way to avoid the arbitrariness of option 1 would be to find one particular, naturally privileged frame of reference. Proposals for singling out such a frame will be discussed in the next two chapters. With option 2 we incur a similar dilemma: if all events in the elsewhen of an event are determinate, then why not the events in their respective elsewhens? Of course, these

---

[26] pp. 140–145.

[27] pp. 148–149.

latter include events in the future light cone of *a*. In other words, given option 2 there appears to be no escape from the Rietdijk-Putnam argument. The only way out would appear to be to arbitrarily privilege *a,* and declare that option 2 applies only to it. Option 3 likewise seems to privilege one frame of reference in an unjustified way. Given it, even events in the remote past would be exempt from determinateness, as long as they are not in *a*'s past light cone. For example, if *a* is the here-now on earth, events in the earth's past would be regarded as determinate, but not those which occurred, say, one million years ago (according to the earth's frame) in the Andromeda galaxy, which is about 2.3 million light-years away. In addition, Putnam points out a further difficulty with option 3:

> Let Oscar be a person whose whole world-line is outside of the light-cone of me-now. Let me-future be a future 'stage' of me such that Oscar is in the lower half of the light-cone of me-future. Then, when that future becomes the present, it will become true that Oscar *existed,* although it will never have had a truth value to say in the present tense "Oscar exists now." Things could come to *have been,* without its ever being true that they *are*![28]

Given that options 1–3 arbitrarily privilege observers in a way which strains credibility, we are left with option 4.

It is often assumed that the existence of a privileged hypersurface of simultaneity is a necessary condition for there to be an open future which can be influenced by agents located in spacetime. However, I will argue that a dynamic view of time can dispense with absolute simultaneity, and also with any notion of purely temporal determinism, that is, of an event's being determinate in virtue of its time of occurrence.

3.5 World Lines, Light Cones, and Endurance

In view of the philosophical explorations on time and causality which will be undertaken in the remainder of this book, a further note on the scope and limits of the Minkowski structure of spacetime is necessary:

---

[28] Putnam (1967), p. 246. Emphases in the original.

All points on the path of an object through spacetime, that is, on its world line, are timelike-related. At any point on this world line, that is, at all stages of its existence, the object has a light cone associated with it, defining the causal past which can affect it, the future it can affect, and the elsewhen which is out of its causal reach. It is important to note that this picture applies only to enduring objects with continuous paths in spacetime. Enduring objects retain their identity throughout the stages of their existence, but can be causally affected and thus undergo change in their properties. Such entities are traditionally known as substances. However, the structure of world lines and light cones breaks down for entities whose identity over time is not defined, as well as for those with discontinuous paths. The path of an electron, for example, allows spacelike jumps according to quantum mechanics. In particular, a bound electron can change its orbit discontinuously. Also, the token identity of an electron over time is not well defined, so that there is no reason why a certain electron on your desk now should not be identical to an electron in the sun a second later.[29] As H. Reichenbach explains:

> An ... essential feature of time is the distinction of a *sense of direction;* and this sense itself is based on the fact that time—and only it—is the dimension of the *lines of action,* upon which we have constructed our teaching of spacetime. Thus, time is the *direction of the fibers* of the manifold, the direction in which the causal chains extend, while space merely contains the bundles of fibers next to each other. The direction of the causal chains is at the same time the direction of the world lines of things remaining identical with themselves, which represent a special case of a causal chain... We call the points of a timelike world line states of *the same thing...* Now, the doctrine of genetic identity has of course suffered a blow through the dissolution of the concept of substance... the choice of the fibers is [therefore] affected by a certain arbitrariness.[30]

[29] In E. J. Lowe's words, "while an electron is *one* entity of a certain kind, there may apparently be no fact of the matter as to *which* electron an electron is and consequently nothing which 'makes it the very electron that it is'." See his article "Individuation", in M. J. Loux, D. W. Zimmerman (eds.) *The Oxford Handbook of Metaphysics* (2003), here p. 78. Emphases in the original.

[30] "Eine ... wesentliche Eigenschaft der Zeit ist die Auszeichnung eines *Richtungssinnes;* und dieser wiederum beruht darauf, dass die Zeit—und nur sie—die Dimension der *Wirkungslinien* ist, auf die wir ja die Raum-Zeit-Lehre aufgebaut haben. So ist die Zeit die *Faserrichtung* der Mannigfaltigkeit, diejenige Richtung, in der sich die Kausalketten erstrecken, während der Raum

To sum up this chapter: it can, I think, be safely concluded that arguments from SR do not succeed in establishing a view of spacetime whereby "everything is written" and future events are unalterable. The main reason is, to quote once again Howard Stein, that "determinism apart from causality is a pseudoproblem". But it is evident that SR does indeed massively challenge the idea of partitioning spacetime into regions of a "settled past" and an "open future", and with it the "naïve" view of time.

---

nur das Nebeneinander der Faserbündel aufnimmt. Die Richtung der Kausalketten ist zugleich die Richtung der Weltlinien von sich identisch bleibenden Dingen, die einen ausgezeichneten Spezialfall einer Kausalkette bedeuten… Die Punkte einer zeitartigen Weltlinie aber nennen wir Zustände *desselben Dinges* … Nun hat freilich die Lehre von der Genidentität eine Erschütterung erfahren durch die Auflösung des Substanzbegriffs… [daher] ist die Wahl der Faserung mit einer gewissen Willkür behaftet." H. Reichenbach, *Philosophie der Raum-Zeit-Lehre* (1977), pp. 317–318. Emphases in the original.

4

Cosmic Time

We have explored the implications of the special theory of relativity (SR) for time by considering frames of reference moving in spacetime. The upshot was that no total ordering can be given for events in spacetime which could be regarded as ontologically privileged in a non-arbitrary way, although events are in general members of some partial temporal ordering. On the other hand, the last century has seen the emergence of an evolutionary cosmology[1] which identifies stages of the development of the universe and puts these stages into a definite order. This presupposes the notion of "the universe at time *t*", which appears at first sight to contrast with the lesson drawn from SR, whereby we cannot assign in an invariant way a stage of the world to a time. The construction of a model of cosmic development in time is of course possible from any frame of reference, although caveats about the limits of such modelling must be borne in mind, as will be laid out in more detail later. Two questions arise in this context: first, whether there is a total order of stages of cosmic development which can be viewed as in some sense privileged over and

[1] On this term, cf. footnote 1 in the introduction.

D. Saudek, *Change, the Arrow of Time, and Divine Eternity in Light of Relativity Theory*, Palgrave Frontiers in Philosophy of Religion,
https://doi.org/10.1007/978-3-030-38411-1_4

above the total orders established from particular frames of reference; second, whether A-theoretical predicates—past, present, and future—can be applied to cosmic times, that is, to "the universe at time *t*", in an observer-independent way. I will argue that the first question can be answered in the affirmative, so that the development of the cosmos constitutes a B-series, but that the second must be answered in the negative.

Of course, in this context the crucial distinction between "the universe" and "the observable universe" must be borne in mind.[2] While a theoretically informed, and therefore non-arbitrary discourse about the universe as a whole is possible, this discourse is in principle limited by the fact that our observations only span a tiny part of the cosmos.

The following discussion is based on results from the general theory of relativity and cosmology, a rough sketch of which is provided in Appendix B.

4.1 Defining Cosmic Time

Cosmic time can be viewed as a *presupposition* of cosmology, insofar as the modelling of the cosmos requires it. This will be true if, as Kant held, phenomena in general simply cannot be understood apart from time.[3] Nothing is prejudiced by this point as to whether a mind-independent temporal ordering in fact exists, or as to whether, in a relativistic context, there is an ordering of events which is in some way ontologically privileged over the orderings observed according to particular frames of reference.[4] Furthermore, the presupposition of the notion of "the universe at time *t*" is necessary for formulating the cosmological principle in its present form, according to which "the general overall features of the universe are the same relative to any local observer at corresponding epochs in the evolution of the universe".[5]

---

[2] Cf. G. F. R. Ellis, "Issues in the Philosophy of Cosmology" (2008), p. 1. http://arxiv.org/pdf/astro-ph/0602280v2.pdf.

[3] See *Kritik der reinen Vernunft*, 2nd edn (1787), AA 57–8.

[4] Cf. M. Dorato, *Time and Reality* (1995), p. 191.

[5] R. A. Mould, *Basic Relativity* (1994), p. 393.

Similar philosophical assumptions are employed in standard cosmology: first, the assumption of isotropy, whereby no spatial direction is different from any other, so that the universe looks the same in all directions. If isotropy is true for more than three observers, the universe must also be homogenous in space. Both isotropy and homogeneity are philosophically plausible, and by now empirically well corroborated.[6] From the homogeneity of the universe, it follows that space must everywhere, at a very large scale,[7] have the same curvature. In addition, since space is locally flat,[8] the metric of such a space must reduce to the Minkowski-metric at short distances from any chosen coordinate origin. These requirements lead to the Robertson-Walker metric,

$$ds^2 = Z^2\left[\frac{d\rho^2}{1-\kappa\rho^2} + \rho^2 d\Omega^2\right] - d\tau^2.$$

In other words, this metric can be used by an observer who verifies isotropy and around whom space is homogenous. Such an observer follows the average motion of matter in spacetime and is therefore known as a comoving observer. By contrast, observers in motion relative to the average motion of matter will observe a systematic anisotropy, because the direction into which they are moving will be blue-shifted, and the opposite direction red-shifted. This is in fact the case for earth's frame of reference, since we are moving at about one-thousandth of the speed of light relative to the cosmic rest frame.[9]

The above metric, when introduced into Einstein's field equations, leads to the Friedmann equations:[10]

[6] Ellis (2008), pp. 24–25.

[7] Observational evidence indicates that space is homogenous from a scale of about 100 Mega-parsecs. For comparison, the Andromeda galaxy is located at a distance of about 0.75 Mega-parsecs.

[8] More precisely, if any point is chosen as the origin of a spacetime coordinate system, space is flat at that point, up to the second derivative of the metric tensor.

[9] T. P. Cheng, *Relativity, Gravitation and Cosmology*, 2nd edn (2010), p. 231.

[10] See the end of Appendix B for an explanation of the symbols.

$$2Z\ddot{Z} + \dot{Z}^2 + \kappa c^2 - c^2\Lambda Z^2 = -kc^2 pZ^2$$
$$3\left(\dot{Z}^2 + \kappa c^2\right) - c^2\Lambda Z^2 = kc^4 \mu Z^2.$$

In the solution of these equations known as the Einstein universe, the universe was assumed not to change over time, so that all time derivatives of the scale factor Z were zero. This turned out not to be a viable model since, on the one hand, it could be shown that the Einstein universe was unstable and would either recollapse or begin a runaway expansion, and on the other, Hubble's observations showed that the redshift of galaxies increased with distance, leading to the conclusion that galaxies are receding from each other. In universes exhibiting a change in the scale factor, the Friedmann equations can be solved for t, that is, the time measured by any comoving observer. A simple and noteworthy solution is arrived at by setting the pressure equal to zero (a reasonable approximation, given the small density of matter in space), assuming that space is flat so that κ, the sign of the curvature, is equal to zero, and also setting the cosmological constant Λ equal to zero. Then, every comoving observer will conclude that[11]

$$t = \frac{2}{3H},$$

where it is assumed that, as t goes to zero, so does Z. The present age of the universe is, according to this model, inversely proportional to the measured value of Hubble's constant H at present.

The idea of introducing cosmic time as the time measured by the comoving observer was introduced by Hermann Weyl. In this approach, the average four-velocity[12] of matter in a large portion of spacetime is assessed. This four-velocity is then assigned to a "fundamental particle" located in the portion. Then, in the frame of reference of the fundamental particle, any hyperplane of simultaneity given by t = *constant* will be characterized by isotropy and a homogenous distribution of matter in the cosmic vicinity of the particle. The situation is similar to that of a static

[11] Mould (1994), p. 407.

[12] That is, roughly speaking, velocity through spacetime.

fluid, where molecules are in relative motion, but larger portions of the fluid are at rest. Moreover, for a given stage of cosmic development, characterized by certain parameters such as the value of density of the cosmic fluid, of pressure or curvature, different fundamental observers will agree which values correspond to which times. Since cosmic time in this approach is arrived at by an averaging procedure, Gerald J. Whitrow has observed: "cosmic time is essentially a statistical concept, like the temperature of a gas."[13]

Whitrow also provides an alternative approach to showing that there is a cosmic time, that is, that time passes at the same rate for all observers associated with fundamental particles: He first proves that two clocks travelling through spacetime at a constant distance from each other will measure the same time, given the constant speed of light. More precisely, the time assigned by one observer to an event happening at the spatial location of another observer will coincide with the proper time as measured by the latter.[14] Of course, fundamental particles do not appear to remain a constant distance apart, but rather to recede from each other, given the expansion of the universe. However, a fundamental observer may, in Whitrow's words, "choose a scale of length measurement so that, if he wishes, he can regard this space as static". The distance from one fundamental particle to another can be defined in terms of the parameter $\rho \equiv r/Z$, where r is conventional distance and Z is the scale factor. In terms of ρ, distances between fundamental observers are constant. As Arthur Eddington put it, "the theory of the 'expanding universe' might also be called the theory of the 'shrinking atom'." Given this, it is possible to construct a time function such that a fundamental observer spatially separated from an event will assign the same time to it as another fundamental observer located at that event. Whitrow concludes: "We have found that the relational concept of the universe, according to which there is no independent spatial background against which systematic changes in the geometrical structure of the universe occur, implies the existence of cosmic time."[15]

---

[13] *The Natural Philosophy of Time* (1963), p. 246.

[14] Whitrow (1963), pp. 183–192.

[15] pp. 248–250, which see also for the quote from Eddington. Whitrow uses the variable r also for what is here termed ρ.

4.2 Scope and Limits of Cosmic Time Functions

Our first approach to cosmic time used the assumption of an everywhere homogenous and isotropic universe, which led to the Robertson-Walker metric. Hence, such cosmic time functions are limited by the domain of applicability of these assumptions. Whitrow's approach relied on the availability of a scale of measurement such that points—and the clocks situated at them—will, on cosmic scales, remain a fixed distance apart. This will only be true for all points in a spacetime if the universe expands everywhere at a uniform rate, an assumption which is well corroborated observationally and leads to Hubble's law:

$$ds / dt = H \cdot s,$$

where s is the distance between two galaxies, and H is constant at a given cosmic time, although its value varies with cosmic time. There is uniform expansion exactly if fundamental observers verify the same large-scale properties of the universe at corresponding proper times.

Likewise, the assumptions of homogeneity and isotropy everywhere have strong observational support for the observable universe. What about the domain beyond it? In principle, for any cosmic time t, only statements about events within our particle horizon at t can be tested observationally. Particle horizons are, in Dorato's words, "hypersurfaces separating *at cosmic time t* the observable from the non-observable with respect to a fundamental particle *A*".[16] This is distinct from the visual horizon which is the boundary within which electromagnetic signals can reach us. The latter is within the particle horizon, given that the universe only became transparent to photons some $4 \cdot 10^5$ years after the Big Bang, whereas neutrinos and gravitational waves can reach us from earlier epochs.[17] Matter entering our particle horizon, and subsequently our visual horizon, is observed to be in thermal equilibrium with, and to be

[16] Dorato (1995), p. 200, emphasis in the original. Alternatively, particle horizons can be described as "the limit of matter that we can have had any causal contact with since the start of the universe" (Ellis [2008], p. 14).

[17] Ellis (2008), pp. 14–15 and pp. 26–27.

characterized by the same density and pressure as the matter already within the horizon previously. This supports the inference that homogeneity is also given for regions adjacent to the limits of the observable universe. Nothing can be said with certainty about the vast and possibly infinite further removed regions,[18] and hence about the applicability of cosmic time to them.

The observed thermal equilibrium with causally separated regions—known as the horizon problem—has been explained through the hypothesis of a primeaval inflation. According to it, matter which in the very early universe was in thermal contact was then flung apart at velocities vastly greater than the speed of light. After the expansion slowed down, light could catch up and once again make contact between some previously separated regions. On the assumption that all matter uniformly underwent inflation, it would be homogenously distributed everywhere. In this case, not only would there be hypersurfaces of t = *constant* extending throughout our particle horizon at a given time, but these hypersurfaces would extend throughout the entire universe. For then, if *A* is a comoving observer at our location in spacetime, *B* is one far removed, but still in our particle horizon, and *C* a further comoving observer within *B*'s horizon but outside ours, then *B* will agree with both *A* and *C* about which stages in the evolution of the cosmos correspond to which cosmic times. Since the relation "occurring at the same cosmic time" between events is transitive, every point in spacetime (though only on a large scale) could then be assigned a cosmic time. This, of course, is a theoretical extrapolation which cannot in principle ever be verified, since there are regions of the universe forever beyond the observational reach of any comoving observer: those beyond her event horizon, which is defined as the boundary outside of which no light can *ever* reach her.[19] It could however be falsified by observations of matter coming over our particle horizon not in thermal equilibrium with the matter already there. It must also be

---

[18] Spatial infinity is often assumed for flat and open universes (i.e. for κ equal to zero or -1). The notion of physical infinities is of course beset with difficult problems. For a discussion, see Ellis (2008), pp. 47–48, and R. Rucker, *Infinity and the Mind* (1982), pp. 9–35.

[19] L. Bergström, A. Goobar, *Cosmology and Particle Physics* (1999), p. 65.

added that some versions of inflation theory predict a non-uniform inflation leading to spatiotemporally disconnected "bubbles" of spacetime.[20]

As is well known, there are also event horizons lying within the observable universe: those of black holes. The event horizon surrounding a black hole is constituted by the points such that light signals fired tangentially will remain in orbit around it. The future light cone of an event within the horizon is warped in such a way that no signal can ever escape the black hole.[21] This puts another important limit on cosmic time functions, since there will be no coordinate observers able to put all events in spacetime into a total temporal ordering.

Furthermore, it is of philosophical importance that the existence of cosmic time functions depends on the distribution of matter in the universe. This has already become clear by the above discussion, but the point is illustrated especially forcefully by the exotic solutions of the Einstein Field Equations discovered by Kurt Gödel.[22] A universe described by these solutions is everywhere homogenous, but not isotropic, since galaxies rotate everywhere around a local compass of inertia. This causes the light cones of galaxies at a separation greater than a certain critical radius to tip over in such a way that their world lines close upon themselves. What is more, closed timelike curves pass through every point in such a universe. It has been subject to debate whether these models, while allowed by the Einstein Field equations, are metaphysically possible since, on a closed timelike curve, any event is both in the future and in the past of another event. Some have therefore concluded that the Gödel universe is "causally pathological" and therefore metaphysically impossible.[23] Gödel himself contended that the causal pathology of such models can be avoided, since

---

[20] Ellis (2008), pp. 15–16.

[21] See, for example, Mould (1994), pp. 349–377; G. F. R. Ellis, R. M. Williams, *Flat and Curved Space-Times* (2000), ch. 6.

[22] "An example of a new type of cosmological solutions of Einstein's field equations of gravitation" (1990). Cf. also S. W. Hawking, G. F. R. Ellis, *The Large Scale Structure of Space-Time* (1973), pp. 168–170. For a discussion of the philosophical implications of the Gödel universe, see especially M. Dorato, "On becoming, cosmic time and rotating universes", in C. Callender (ed.) *Time, Reality and Experience* (2002); P. Yourgrau, *The Disappearance of Time: Kurt Gödel and the Idealistic Tradition in Philosophy* (1991); and B. Dainton, *Time and Space*, 2nd edn (2010), pp. 381–386.

[23] Cf. M. Tooley, *Time, Tense, and Causation* (1997), pp. 290–291. See also Dorato (1995), p. 191, footnote 9 for references.

the temporal perimeter is of a dimension which prevents any trace of a past event from being left when an observer following the closed world line returns to it.[24] The backwards time travel would therefore be causally innocuous, and no inhabitants of the Gödel universe would be able to kill their own ancestors before being born.[25] We will revisit the question whether a Gödel universe is possible in Sect. 11.3 of this book.

Gödel himself concluded from the possibility of a universe admitting no cosmic time function that change cannot be objectively real. Dorato rightly questions this jump from a possible world model to a conclusion about time in our universe.[26] If a necessary condition for the existence of an objective lapse of time is that spacetime be such as to allow cosmic time functions, then this condition will be met as long as the factually existing universe is so structured, regardless of whatever different universes may be conceivable. Gödel's argument—provided of course that his rotating universe cannot be ruled out on logical grounds in the first place—does however speak against a strongly substantivalist view of time, according to which the passage of time is a basic feature of reality independent of the content of the world,[27] as is the case with those versions of A-theory which see it as "necessarily true, if true at all".[28]

4.3 Assessing the Significance of Cosmic Time Functions for the Philosophy of Time

Is cosmic time, in a universe where it exists, privileged in any sense? In a certain respect it arguably is: At least in the homogenous and isotropic observable universe, cosmic time allows assigning unique values for certain variables describing the large-scale structure of the universe to points

[24] K. Gödel, "A remark about the relationship between relativity theory and idealistic philosophy" (1949).

[25] On the question of whether time travel can be causally harmless, cf. Dainton (2010), ch. 8.

[26] Dorato (2002), pp. 259–260.

[27] Cf. Sect. 1.1 on the notion of "strong" substantivalism about time.

[28] D. W. Zimmerman, "Presentism and the space-time manifold" (2013), p. 110.

in time. Examples include the values of pressure, matter and radiation density, curvature, Hubble's constant, or of parameters such as Ω_{matter} and Ω_{Λ}.[29] Fundamental observers will agree on which values of these variables correspond to which points in time. Cosmic time alone permits such a simple functional relationship. If simultaneity is defined in any other way, for example, through the hyperplanes of simultaneity of any frame of reference,[30] the values of these parameters will vary, also on a large scale. This means that cosmic time can be used to give a unified, causal account of the evolution of the domain it applies to, in such a way that statements can be made which causally relate later stages of the universe to earlier ones. For example, there are now good theoretical reasons to assume that the universe began to accelerate, after a previous deceleration phase, about seven billion years ago, and that the contribution of the cosmological constant to the mass-energy of the universe began to outweigh that of matter about four billion years ago. This led to a further decrease in density, and also to a stunting of the growth of galaxies in the comparatively recent cosmic history.[31] A further important example is the decoupling of matter and radiation which took place some four hundred thousand years after the Big Bang, whereupon the universe became highly transparent to electromagnetic radiation.

In scientific cosmology, such statements about the evolution of the observable universe, and about stages thereof at different cosmic times, are made routinely. As regards the very early universe, they face the difficulty that the physics at those epochs did not permit systems which could be used as clocks, so that a cosmic time for these stages of cosmic development cannot be understood as the time measured by a comoving observer.[32] Similar caveats apply to the far-future universe.

---

[29] That is, the fractional contribution of matter and cosmological constant, respectively, to the total mass-energy of the universe.

[30] Including that of a comoving observer, since it is not her hyperplane of simultaneity which defines the universe at cosmic time t = *constant*, but rather the atlas consisting of the hyperplanes of simultaneity of all comoving observers.

[31] Cheng (2010), p. 264.

[32] For a more detailed discussion, see S. E. Rugh, H. Zinkernagel, "On the physical basis of cosmic time".

These limitations notwithstanding, cosmic time is arguably the most natural time parameter available for constructing a total ordering of stages of the development of at least the observable universe, since it enables correlating large-scale properties to time as measured by fundamental observers. The set of hypersurfaces described by $t_{cosmic} = constant$ then constitutes a B-series isomorphic to the line of real numbers.

Does cosmic time also constitute an A-series, that is, can the changing A-theoretical predicates past, present, and future be assigned to such hypersurfaces in a mind-independent way? Some authors, whose proposals will be discussed in the next chapter, take this route in order to defend the passage of time against the threat from a timeless block universe.

According to this picture, the cosmic now is the true now, separating the determinate past from the open future. Likewise, every past spacelike hypersurface defined by $t_{cosmic} = constant$ was once the present moment at which the possible became actual throughout the universe. In our region of space, the true now would then be defined by successive hyperplanes of simultaneity of the local comoving frame. Moreover, authors defending the existence of an A-theoretically understood cosmic now in general wish to save the characteristic asymmetry of human agency, by which we act now in order to influence the still open future. It is therefore required that the now-hyperplane of the comoving observer intersect the origin of a spacetime diagram centred on our here-now, lest we earthlings tragically end up in the fixed past or, *per absurdum*, in the not yet existing future. Likewise, for all observers existing now, their future light cones will have to lie entirely in the domain of $t > t_{now}$ in terms of cosmic time, their absolute pasts in $t < t_{now}$, and their hyperplanes of simultaneity will have to intersect at their respective origins with the assumed cosmic hypersurface of simultaneity, but will in general not coincide with it.

Dorato suggests that an A-theoretic interpretation of cosmic time may be viable, emphasizing its proximity to time as measured from our frame of reference:

> Consider now that since we *do see* isotropic expansion in all directions of space, the earth is nearly comoving ... Knowing that the proper time of an observer in a comoving frame can be made common to all points of the manifold, observers on earth can claim objectivity of becoming because

> *every comoving observer would see the same order or succession of states that we see.* This argument can justify the evolutionary talk now prevalent in cosmology, since the possibility that the future domain of dependence of an hypersurface can contain indeterminate events means that a condition that was regarded as necessary for the mind-independence of becoming and the A theory of time would not be violated. With an ironic twist, observers on earth can claim objectivity to their nows if and only if they have no privileged position in the universe.[33]

It seems to me, however, that there are strong reasons to doubt that hypersurfaces of cosmic simultaneity can be assigned A-theoretical predicates: In order for there to be an ontologically significant division across spacetime between past and future events, it is necessary that there be a total ordering of temporal precedence on the set of all events—for only under this condition do all events happening simultaneously form an equivalence class, which can be thought of as a hypersurface of simultaneity, and one of these classes would contain all events happening "now". We have seen above that the statistical averaging procedure used to construct cosmic time functions does indeed allow the construction of a total ordering of *stages* of the universe in a unique way. This, however, does not amount to a total order of all *events*, because it remains true that clocks tick in accordance with their state of motion and the gravitational potential in which they are situated. In particular, cosmic time does not change the fact that no invariant ordering for any two spacelike-separated events is available. It remains true, therefore, that there is no "set of all events occurring at a given time" and, *a fortiori*, no "set of all events occurring now". But then, there is also no global past-present-future distinction. Thus, evolutionary cosmology does not enable us to return to the view of a global passage of time.

Furthermore, there is no indication from any known physical theories or results that the passage of time is responsible for any determining of the indeterminate. To assume that it does leads not only to the difficulties besetting A-theory in general but also to problems such as double determination, as we will see in Chap. 6, where I will argue that the conversion

---

[33] Dorato (1995), p. 208, emphases in the original. However, Dorato does not commit himself to an A-theoretic reading of cosmic time.

of the possible to the actual cannot in principle be understood in terms of physical law. Moreover, the known physical laws are time-symmetric, and in particular no time asymmetry is associated with the characteristic indeterminism of microphysics. Unidirectionality of time, and the action asymmetry by which we can only act towards the future, whereas we cannot change the past, must indeed be taken seriously, but in my view these aspects of reality can be better accounted for through a theory of time based on local causal interactions, without the need to posit the existence of a cosmic now-hypersurface in order to save the phenomena.[34]

4.4 Does Quantum Mechanics Save Cosmic Simultaneity?

However, an important argument in favour of absolute simultaneity must still be considered, namely that from entangled states in quantum mechanics. As is well known, this phenomenon raises many profound questions in the philosophy of physics, but I will limit the discussion to its bearing on the issue of simultaneity.

Measurements can be made on quantum mechanical systems—for example, microscopic particles—in order to get information about *observables*. These include, among others, the position, momentum, spin or energy of a particle. However, multifarious experimental evidence—such as the well-known double-slit experiments, and experiments involving measurements of the spins of particles—indicates that quantum mechanical systems do not actually possess such observable properties unless they are measured.[35] Rather, according to the theory derived from such experiments, a quantum mechanical system is described by a state vector Ψ, which can be constructed in the following way: let q designate a measurable value of a particular observable, and let χ_q denote the state of a system exhibiting this value, known as an eigenstate. Each χ_q has an

[34] For a critique of attempts to reconstruct a worldwide now from cosmology, see also R. J. Russell, *Time in Eternity* (2012), pp. 309–313.

[35] See, for example, M. A. Morrison, *Understanding Quantum Physics* (1990), pp. 18–96; D. Z. Albert, *Quantum Mechanics and Experience* (1992), pp. 1–72.

associated probability amplitude c_q, a complex number whose square modulus—which is a real number—gives the probability that the system, upon measurement, will exhibit the value q. The state vector is then written as the sum of the eigenstates multiplied by their probability amplitudes[36]:

$$\Psi = \sum_q c_q \chi_q.$$

That is, if a quantum mechanical system is not measured, it does not exist in a definite state with definite values for its observables, but instead is in a state of *superposition*, characterized only by *probabilities* for being found to be in its particular eigenstates.[37]

In some cases, a system of *several* particles is described by a single state vector which cannot be decomposed into state vectors of the particles which are part of the system. The states of such systems are therefore known as non-separable, or entangled states. For example, it is possible to prepare two electrons in such a way that the spin-state of one is entangled with that of the other. If the spin of one of these electrons is measured, the joint wave function collapses into an eigenstate, thereby determining also the spin of the other. Suppose, for example, that A and B are two entangled electrons and that we measure the spin of A in the z-direction according to some coordinate system. If A's spin is measured to point in the positive z-direction, that is, if A is "z-spin up", B must be "z-spin down" and vice versa. A and B can be separated by an arbitrarily large distance. Since neither has a definite spin before being measured, and since the z-spin value of one immediately determines that of the other, we are faced with what seems like an instantaneous interaction across space. Experiments of this type were first proposed by Albert Einstein, Boris

[36] Morrison (1990), pp. 468–472. Some observables have continuous, rather than discrete eigenvalues, so that the state vector can also be written as an integral of the eigenstates of such an observable, with each one multiplied by its respective probability amplitude.

[37] As is well known, other interpretations than the Copenhagen interpretation, which I am assuming here, have been advanced. I will leave these aside in the current discussion, since the "orthodox" interpretation, for all we currently know, best agrees with experiment.

Podolsky and Nathan Rosen in order to show that quantum mechanics cannot be a complete theory—hence the name "EPR-experiments".

These findings seem to indicate the existence of an absolute simultaneity: the event of *A*'s acquiring a definite state for a particular observable (in our example, *z*-spin) is simultaneous with *B*'s acquiring a corresponding state (the opposite *z*-spin).[38] It has been argued that this establishes a unique, true reference frame, for it seems that only by using the frame according to which these two spacelike-separated events are simultaneous do we get the correct account of what is going on, whereas according to any other frame, one of the two events would be viewed, incorrectly, as occurring before the other. Some authors, whose work will be discussed in more detail in the next chapter, have proposed that this "true" simultaneity, apparently revealed by quantum mechanics, points to the existence of hypersurfaces of cosmic simultaneity. To use the words of Howard Stein in his criticism of such a model put forward by Nicholas Maxwell: "Maxwell proposes … that there is a unique globally defined system of space-like hypersurfaces that partition space-time into the loci of all possible such collapses. This system, then, would constitute his re-established objective simultaneity-relation."[39] According to some such proposals, to be discussed in the next chapter, simultaneity according to the "true frame" supposedly established by measurements on entangled states coincides with simultaneity according to hypersurfaces of $t = constant$ established by cosmic time functions.

However, there are strong reasons against the idea that wave function collapse in entangled particles occurs according to a privileged foliation of spacetime established by cosmic time functions, or indeed according to any other such foliations. For an EPR-experiment can be conducted in any frame of reference, and there is no reason to suppose that the hypersurface of simultaneity containing the two spacelike-separated events discussed above should coincide with that established by the assumed privileged foliation. Suppose, for example, that an EPR-experiment is

[38] It can rightly be questioned whether we are really faced with two events here, rather than a single one, namely the collapse of the joint wave function describing the state of both entangled particles. I am using the talk of *two* events merely because this is necessary in order to present the argument from entangled states to absolute simultaneity between spacelike-separated events.

[39] "On relativity theory and openness of the future", *Philosophy of Science* (1991), here p. 165.

conducted in some frame of reference which does not coincide locally with the postulated privileged frame. By hypothesis, the two events are simultaneous according to the latter frame. They are therefore not simultaneous in the former, so that, according to this frame, there will be a time interval in which one of the two particles, say *A*, has acquired a definite spin, whereas *B* has not. If the spin of *B* is then measured during this time interval, it is quite possible for it to have the same spin as *A*, leading to a violation of conservation of angular momentum. But this contradicts the theory whose truth we are assuming, that is, quantum mechanics.

Craig Callender has similarly pointed out that the assumption of a preferred frame conflicts with quantum mechanics. His argument relies on the Bohmian interpretation of quantum mechanics, rather than the standard interpretation which I am assuming. Nevertheless, the upshot is a similar one:

> There is an in principle irresolvable *coordination problem* between the two preferred foliations, the metaphysically preferred foliation posited by the tenser and the physically preferred one by Bohmian mechanics. There is simply no reason to think the two are the same. Only blind faith leads one to expect that the two are coordinated. In our above experiment, A might measure first and then B measure second according to the Bohm frame, yet according to the temporal becoming frame B measures first and A second. Assuming the becoming frame is primary, we would say B really happened before A; meanwhile fundamental physics would say that A happened before B. Since it would be a miracle if the two frames coincided exactly, with near certainty this will be the case for *some* pairs of events. *Hence the tenser is committed to asserting that with near certainty fundamental physics gets the order of some events the wrong way round.*[40]

In sum, quantum mechanics does not corroborate the case for the existence of frame-independent, absolute simultaneity and a preferred foliation of spacetime, but rather undermines these notions still further.

[40] "Finding 'real' time in quantum mechanics" (2007), p. 13. http://philsci-archive.pitt.edu/4262/. Emphases in the original.

5

Worldwide Becoming, Block Universe, or Some *Tertium*? Contemporary Answers to the Problem of Time and Relativity Theory

As has become clear in the preceding chapters, special and general relativity theory pose a double challenge to the philosophy of science: on the one hand, we need to "save the phenomena" and account for the familiar and intuitively obvious features of time, in particular its irretrievable flow from the future via the present to the past, or at least explain how the overwhelming impression arises that time has these features. On the other hand, this objective must be met without invoking any implausible physics in order to defend flowing time and to rescue absolute simultaneity from relativity theory. In this chapter, we will consider in greater detail different answers to this problem. Of course, it is not possible to give a comprehensive overview of the relevant literature, so I will concentrate on a few authors representing the principal options for addressing the problem of time and relativity theory: cosmic A-theory, the block universe model, and proposals towards a third option which incorporates elements of the other two.

D. Saudek, *Change, the Arrow of Time, and Divine Eternity in Light of Relativity Theory*, Palgrave Frontiers in Philosophy of Religion,
https://doi.org/10.1007/978-3-030-38411-1_5

5.1 Cosmic A-theory—in Search of the True Present Hypersurface

A-theorists assume a mind-independent passage of time from the future via the present into the past. In the context of spacetime physics, it is therefore necessary to defend the existence of absolute simultaneity in such a way that, for every moment in time, there is a hypersurface—which need not necessarily be a hyperplane, that is, a flat hypersurface—containing all events occurring at that moment, whose content is independent of the frame of reference. In particular, one such hypersurface defines the worldwide "now" which separates the past from the future and which can be thought of as a global "wave" of becoming. On the crest of this wave, the open future turns into the fixed past. I will use the term "cosmic A-theory" to refer to this view of time. To identify the necessary privileged hypersurfaces, cosmic A-theorists need to posit additional spacetime structure than that afforded by relativity theory. Several scholars, whose work is cited below, have advanced detailed proposals to that end.

An argumentative cornerstone of cosmic A-theory is the passage of time as experienced, with its intuitively obvious distinction between past, present, and future. Since this cannot be simply dismissed as illusory, proponents of this theory argue that there must be a global present. To quote the physicist Lee Smolin:

> We have direct experience of the world in the present moment … it is … an undeniable feature of the natural world that qualia are experienced in moments which are experienced one at a time. This gives a privileged status to each moment of time, associated with each experience … This means that we have direct access to a feature of the presently present moment … Any description of nature that does not allow *now* to be intrinsically defined is an incomplete description of nature.[1]

[1] R. M. Unger, L. Smolin, *The Singular Universe and the Reality of Time: A Proposal in Natural Philosophy* (2015), pp. 481–2.

In a similar vein, the philosopher Dean W. Zimmerman, in his paper "Presentism and the space-time manifold",[2] has presented an argument which starts from the experience of the passage of time and concludes that global hypersurfaces of simultaneity must exist.[3]

How does the global passage of time in cosmic A-theory square with the lack of a frame-independent simultaneity in spacetime physics? Proponents of the theory have come up with several ideas for finding a physical basis for global time in spite of relativity theory. These can be divided into three groups:

The first is to use the cosmic time functions which emerge from general relativity theory given the fact that we are living in a homogenous and isotropic universe, as discussed in detail in Chap. 4. Authors such as William L. Craig,[4] as well as Dean W. Zimmerman,[5] argue that cosmic time is suited to provide the privileged frame needed for the universal past-present-future distinction. On this basis, Craig advances the case that cosmic time coincides with metaphysical time, which is also the time which God experiences. He corroborates his case for a privileged frame of reference by identifying two further "modern equivalents of the aether" suited to define it: the microwave background radiation, and the quantum mechanical vacuum.[6] Roberto M. Unger and Lee Smolin likewise use cosmology as an argument for a time which is "non-emergent, global, irreversible, and continuous"—indeed, it is so basic that it remains "when everything else leaves".[7] According to them, cosmic time constitutes a total order of all events, rather than only of large-scale stages of the universe: "any event, anywhere in the universe, fits on a single cosmological time line."[8] Also, it is not just a B-series, but rather grounds mind-independent A-theoretical properties of states of the universe and of

---

[2] In C. Callender (ed.) *The Oxford Handbook of Philosophy of Time* (2013).

[3] pp. 41–52.

[4] See his *Time and the Metaphysics of Relativity* (2001), especially ch. 10.

[5] *Op. cit.*, pp. 71–90.

[6] *Op. cit.*, pp. 219–222.

[7] *Op. cit.*, pp. 226–7.

[8] p. 214.

events: there is a present moment of the entire universe, as well as a concomitant global distinction between past and future.[9]

The second strategy for obtaining absolute simultaneity and hence, global time, from physics is based on quantum mechanics, in particular from EPR-type experiments, as discussed in Sect. 4.4. This move is endorsed by Dean W. Zimmerman,[10] by William L. Craig[11] and—as we have seen in Sect. 2.2.5—by John R. Lucas. According to these accounts, the hypersurface along which the collapse of two spatially separated entangled particles occurs defines a privileged standard of simultaneity. In Craig's thought, quantum mechanical simultaneity furthermore coincides with simultaneity as defined by the cosmic frame. He quotes the physicist Antony Valentini in support, according to whom

> … the absolute 3-space along which subquantum nonlocality acts … will necessarily coincide with the observed rest-frame defined by the uniform microwave background. We therefore predict that, if the nonlocality ever becomes directly observable, it will be found to propagate along the hypersurface defined by the observed cosmological rest frame.[12]

The third strategy is to use the Neo-Lorentzian interpretation of the special theory of relativity (SR) as a foundation for global simultaneity, a move endorsed by William L. Craig[13] as well as Michael Tooley.[14] This interpretation allows retaining absolute simultaneity in spacetime because, according to it, only the average round-trip speed of light is equal to *c*, but its instantaneous speed depends on the direction it is travelling in. Moreover, according to this account, Lorentz contraction is explained by motion of an object relative to the ether, and is thus accounted for dynamically rather than kinematically.

---

[9] pp. 246–7 and 491.

[10] *Op. cit.*, pp. 80–1.

[11] *Op. cit.*, pp. 223–235.

[12] Quoted on p. 234 n. 110, from A. Valentini "Pilot-wave theory of fields, gravitation and cosmology", in J. T. Cushing, A. Fine, S. Goldstein (eds.) *Bohmian Mechanics and Quantum Theory: An Appraisal*, Dordrecht: Kluwer (1996), pp. 45–66, here p. 63.

[13] *Op. cit.*, ch. 9.

[14] *Time, Tense and Causation* (1997), pp. 339–360.

5.2 The Block Universe

The fact that relativity theory undermines the concepts of a uniquely defined present, past, and future of the universe has led several thinkers to conclude that "everything is written", that the four-dimensional block constituting the entire history of the universe is in a sense "already there" and that therefore the notion of true becoming is illusory. This idea is well illustrated by a quote from Kurt Gödel, who held that relativity of simultaneity leads to a fundamentally changeless world:

> It seems that one obtains an unequivocal proof for the view of those philosophers who, like Parmenides, Kant, and the modern idealists, deny the objectivity of change and consider change as an illusion or an appearance due to our special mode of perception. The argument runs as follows: Change becomes possible only through the lapse of time. The lapse of time, however, means (or at least, is equivalent to the fact) that reality consists of an infinity of layers of "now" which come into existence successively. But, if simultaneity is something relative ... reality cannot be split up into such layers in an objectively determined way. Each observer has his own set of "nows", and none of these various systems of layers can claim the prerogative of representing the objective lapse of time.[15]

The same train of thought appears in the definition of the block universe given by John C. Polkinghorne and Christopher J. Isham:

> The central thesis of those who support the idea of a block universe is that (i) the notion of a set of spacetime points is meaningful; and (ii) all such points have an equal ontological status. In particular, no fundamental meaning is to be ascribed to the concepts of "past", "present", and "future". Thus the notion of an ephemeral "now"—a moving barrier that continuously travels into the future, leaving behind an unchangeable past—is regarded as being a construct of the human mind that has no reference to reality as understood in the framework of modern physics. A secondary implication is that our perception of a genuine openness of future events is

[15] "A remark about the relationship between relativity theory and idealistic philosophy" (1949), here pp. 202–3.

> essentially illusory or, at the very least, needs to be understood within this framework of an "eternally existing" spacetime. Such a view of spatial and temporal relationships can raise serious difficulties for any theological or philosophical position in which the concept of human freedom plays a significant role.[16]

We have already considered the argument for temporal determinism advanced by Wim Rietdijk and Hilary Putnam, according to whom the transitivity of the relation "being present" between events renders the entire content of spacetime fixed and unchangeable. But the block universe model also enjoys the support of some of the founders of modern spacetime physics themselves. For example, as is well known, Albert Einstein believed that the distinction between present, past, and future is illusory, as is illustrated in his writing of condolence to the widow of his friend Michele Besso: "People like us, who believe in physics, know that the distinction between past, present and future is only a stubbornly persistent illusion."[17] As Rudolf Carnap recounts, Einstein was deeply interested in the "now" of human experience, but held that the difference between past, present, and future "does not and cannot occur within physics".[18] Einstein was furthermore a determinist, ruling out the possibility of an open future:

> Everything is determined, the beginning as well as the end, by forces over which we have no control. It is determined for the insect as well as the star. Human beings, vegetables, or cosmic dust, we all dance to a mysterious tune, intoned in the distance by an invisible piper.[19]

While Einstein's views on time and causality imply a block universe where "everything is written", it is important to note that Einstein's determinism, as is apparent from the last quote, is causal rather than temporal.

---

[16] C. J. Isham, J. C. Polkinghorne, "The debate over the block universe", in R. J. Russell, N. Murphy, C. J. Isham (eds.) *Quantum Cosmology and the Laws of Nature* (1999), here p. 139.

[17] S. Hawking, *A Stubbornly Persistent Illusion: The Essential Scientific Works of Albert Einstein* (2009), back cover.

[18] P. A. Schilpp, *The Philosophy of Rudolf Carnap* (1963), pp. 37–8.

[19] R. W. Clark, *Einstein: The Life and Times* (1979), p. 422.

The view that future events are "already there" has also been endorsed by the early relativist Arthur Eddington, who famously stated:

> The division into past and future (a feature of time-order which has no analogy in space-order) is closely associated with our ideas of causation and free will. In a perfectly determinate scheme the past and future may be regarded as lying mapped out—as much available to present exploration as the distant parts of space. Events do not happen; they are just there, and we come across them. "The formality of taking place" is merely the indication that the observer has on his voyage of exploration passed into the absolute future of the event in question.

This means that determinate events, linked by deterministic causal laws, are spread out throughout spacetime:

> A detached observer contemplating our world would see some events apparently causing events in their future, others apparently causing events in their past—the truth being that all are linked by determinate laws, the so-called causal events being merely conspicuous foci from which the links radiate.[20]

A particularly clear statement in favour of a block universe model is given by Olivier C. de Beauregard in his paper "Time in relativity theory: Arguments for a philosophy of being"[21]:

> There can no longer be any objective and essential (that is, not arbitrary) division of spacetime between "events which have already occurred" and "events which have not yet occurred". There is inherent in this fact a small philosophical revolution… If matter has spatial extension, it follows … that it has also extension in time. This is why first Minkowski, then Einstein, Weyl, Fantappiè, Feynman, and many others have imagined spacetime and its material contents as spread out in four dimensions. For those authors, of whom I am one, who take seriously the requirement of covariance, relativity is a theory in which everything is "written" and where

[20] A. Eddington, *Space, Time and Gravitation: An Outline of the General Relativity Theory* (1920), pp. 46–7.

[21] In J. T. Fraser (ed.) *The Voices of Time* (1981), pp. 417–433.

> change is only relative to the perceptual mode of living beings. Humans and other living creatures, for reasons which one can try to explain, are compelled to explore little by little the content of the fourth dimension, as each one traverses, without stopping or turning back, a time-like trajectory in spacetime…
>
> There are writers who affirm that the future contains elements which are undetermined by the past and the present, and that the future light cone in the diagram must therefore be left blank. The answer is that there *is* a future. Nature "will take" one of the alternatives open to her, and it is this that we must imagine inscribed, even though we do not know what it "will be".[22]

Beauregard is of course aware that this picture conflicts with the intuitions whereby time is a one-way street, and whereby furthermore we can act freely to influence the future. He explains the former phenomenon in terms of irreversible processes in nature, such as increase of entropy at macroscopic levels, and emission of radiation by charged particles, phenomena which are themselves probabilistic in nature: "the irreversible evolution of the universe imposes, *via* probability, its dissymmetry even on the microscopic scale."[23]

While Beauregard denies the possibility of an indeterminate future light cone, because its content is "written",[24] he does not therefore squarely deny free will, but rather argues for a compatibilist picture, whereby the history of the universe can be recounted in two ways: First, in terms of the "psi-function of the entire universe", which

> denotes which possible states are occupied and which are not. In the Heisenberg picture the occupation numbers of the states never change: nothing happens. This picture is a reformulation in quantum terms of the classical and deterministic conception of nature in which everything is written once and for all.

---

[22] pp. 429–430, emphasis in the original.

[23] p. 431. Beauregard undertakes a detailed discussion of the relationship between thermodynamics, radiation emission and absorption, probability and the passage of time in "La grandeur physique temps", in J. Piaget (ed.) *Logique et Connaissance Scientifique* (1967).

[24] "It seems to be urgently implied that the future half of the Minkowski diagram cannot be filled in, even in principle, because decisions upon which it depends have not yet been made. To this we have answered above that *something* is going to happen and that is what is written down." "Time in Relativity Theory", p. 432.

The other way is the "interaction picture" for which there is some indeterminacy. However, according to Beauregard, the effect of this indeterminacy consists in the fact that

> our successive sense impressions cannot in general be put in exact causal relation with each other ... in this sense, a four-dimensional diagram of an observer's sensory experience will not consist of causally related events.[25]

In the picture drawn by Beauregard, these two views are complementary rather than mutually exclusive. However, the indeterminacy in the second view applies only to experience, not to action, so that there is no notion of a "branching" future which would enable choices by agents to actually change the content of spacetime. In this sense, Beauregard's view is similar to the compatibilism about free will advanced by Immanuel Kant.

5.3 Why Neither Cosmic A-theory nor the Block Universe Works

Neither cosmic A-theory nor the block universe lead to a satisfactory account of the passage of time in the context of relativity theory. Cosmic A-theorists rightly emphasize that a viable theory of time must account for its passage and its asymmetric character, for the irretrievability of the past and the openness of the future, rather than simply declaring these features to be unreal by spatializing the time axis, and by appealing to the relativity of simultaneity. However, proponents of the theory assume that true becoming is possible only if there is an absolute flow of time with a global present, and in this way, extrapolate from an indeed central feature of our experience to the structure of the world as a whole. This leads essentially to a doubling of temporal dimensions, since an absolute time existing over and above the time measured by local clocks is postulated in order to save the global past-present-future distinction. In this picture, time is in general not what the clock measures, unless the clock happens to be carried by a privileged observer. Thus, while the cosmic hyper-time

[25] pp. 432–433.

is meant to save experienced temporal asymmetry, it only leads to the additional problem of how it is that the past-present-future distinction which we experience locally happens to coincide with the cosmic one.

The three proposals advanced by cosmic A-theorists in order to provide a foundation for the global wave of becoming suffer from their sheer physical implausibility:

First, cosmic time functions cannot provide a foundation for a global "now", as can be seen from a simple argument: A global "now" would consist of a particular set comprising all events happening simultaneously. For such a set to exist, a total order of events is necessary, so that for any two events x and y, either x is before y, or the converse, or they are simultaneous. But, as we have seen in Chap. 4, cosmic time functions do not permit all events to be so ordered—unlike maintained by Unger and Smolin—but only large-scale states of the cosmos. Therefore, they do not deliver what the cosmic A-theorist needs.

Second, quantum mechanical simultaneity does not corroborate the case for cosmic simultaneity, as contended by Craig and Zimmerman, but rather weakens it, since there is no reason to suppose that simultaneity as defined through cosmic time functions coincides with that occurring in quantum mechanics. If it did so coincide, violations of the predictions of quantum mechanics would result, for reasons given in Sect. 4.4.

Third, the Neo-Lorentzian view of spacetime rests on implausible physical assumptions: the existence of an ether, and an anisotropy in the speed of light. It is ontologically simpler to do without the former, and to assume that the laws of nature are the same in every frame of reference, so that speed of light is everywhere equal to the fixed value resulting from Maxwell's equations. However, it is also important to note that Neo-Lorentzian theory cannot be conclusively disproved experimentally: to do so would require measuring the one-way speed, as opposed to the round-trip speed, of light. This, however, can only be done by using two spatially separated clocks assumed to be synchronized, an assumption which cannot be taken for granted, because clock transport involves time dilation both in standard relativity theory and in the Neo-Lorentzian theory, according to which moving clocks interact with the ether. Notwithstanding this experimentally open issue, it is preferable to account for time's passage with a less cumbersome ontology, compatible with standard relativity theory, than that proposed by the Neo-Lorentzians. A proposal to this end is offered in Chap. 9.

In sum, cosmic A-theory takes our experience of the passage of time seriously but extrapolates the structure of this experience to the universe as a whole, and thus essentially has the universe comply to our deep-seated pre-relativistic intuitions.

The evaluation of the block universe model is simpler: This model incurs no conflicts with physics in the way in which cosmic A-theory does. Its principal problem is its failure to account for the passage of time as experienced. A fully satisfactory model would need to save the phenomena in some way. If it declares them to be illusory, it would at least need to be shown how the illusion arises. We have seen, in the context of the Rietdijk-Putnam argument discussed in Sect. 3.2, that the characteristic temporal determinism of the block universe fails, principally because the content of the future light cone of a spacetime point depends on what is happening at that point, and can therefore be influenced. Furthermore, the very notion of temporal determinism over and above causation was seen to arise from the confusion between *necessitas consequentis* and *necessitas consequentiae.* By the same token, the lack of a frame-independent past-present-future distinction in relativity cannot be cited in support of the view whereby the four-dimensional spacetime block constituting the universe is "already there" so that "everything is written". Indeed, like the A-theoretic models discussed earlier, this notion itself clandestinely introduces a kind of hyper-time existing externally to the manifold, as has been well put by Huw Price:

> In an attempt to highlight the contrast with the dynamic character of the "moving present" view of time, people sometimes say that the block universe is *static.* This is rather misleading, however, as it suggests that there is a time frame in which the four-dimensional block universe stays the same. There isn't of course. Time is supposed to be included in the block, so it is just as wrong to call it static as it is to call it dynamic or changeable. It isn't any of these things, because it isn't the right sort of entity—it isn't an entity *in* time, in other words.[26]

[26] H. Price, *Time's Arrow and Archimedes' Point: New Directions for the Physics of Time* (1996), pp. 12–13, emphases in the original.

The block universe model thus likewise extrapolates a pre-relativistic notion onto spacetime as a whole, which in this case is the ontological status of the past. Why, one may ask, shouldn't the ontological status of the future, by parity of reasoning, be extrapolated in this way, landing us in a fuzzy block?

The failure of temporal determinism is of course of no avail against Einstein's block universe, since the determinism in it is causal. However, as we will see in the next chapter, it is questionable whether a causally deterministic universe can evolve at all, so that Einstein's block seems to conflict with evolutionary cosmology. Of course, the downfall of determinism in physics furthermore weakens the case for such a universe.

5.4 A Third Way?

At this point, readers may well wonder whether some third way is available which overcomes the dichotomy between cosmic A-theory and the block universe, an alternative which posits no implausible physics but yet is able to save the phenomena of the passage of time as experienced. Some proposals pointing in this direction are indeed available. A brief overview of this literature follows, where I will limit myself to describing some of the key points therein:

First, Robert J. Russell puts forward a "relational and inhomogenous spacetime ontology",[27] compatible with spacetime physics and yet hospitable to flowing time. Russell's crucial move to arrive at such an ontology is to propose that relations between events, rather than properties of events, ground the distinction between past, present, and future. In this way, events are never assigned contradictory properties, so that McTaggart's paradox does not arise. At the same time, the difference between past, present, and future is not merely subjective, being grounded in relations which truly hold between events. The relational structure by which events are connected leads to a complex, fractal structure, whose topology Russell explores in detail.[28]

[27] *Time in Eternity* (2012), pp. 299–304.

[28] *Op. cit.*, chs. 5 and 6.

A further contemporary author who argues that true becoming does not demand a total and global temporal ordering is Steven F. Savitt, who rejects the argument against the passage of time and true becoming advanced by Beauregard by emphasizing two important distinctions: First, that between coordinate time and proper time measured along a world line. And second, between becoming, understood as the "mere occurrence of events at various serially ordered clock times" on the one hand, and tense, with its distinction between present, past, and future on the other. Savitt, citing the support of Adolf Grünbaum, proposes that, while change is mind-independent, tense is mind-dependent. The reason is that the tensed predicates "now", "past", and "future", are indexical terms, and therefore depend on there being language-users to employ them. Furthermore, such predicates do not represent any "physical attribute" in their referents. This distinction between change and tense means that Gödel's argument, whereby relativity of simultaneity and the downfall of a global present also entail a world without temporal passage, is invalid. In Savitt's own words:

> … time *qua* passage is a *local* phenomenon, tied to a world line. For eons we have tied passage to an advancing *global* now, and this idea is buried deep in our worldview. It is an idea that we must transcend.[29]

James Harrington likewise argues that a robust difference between a determinate past and an open future can be upheld in a spacetime which lacks such a distinction globally.[30] His approach is based on the notion of relational indeterminacy—or, conversely, determinacy—which holds between regions of spacetime. In his own words: "One region of spacetime P is relationally indeterminate relative to another region Q when the causal past of Q fails to determine the state of P."[31] Thus, spacetime itself comes with a structure which accommodates a modal distinction: "[T]he past null-cone and its interior of each point of Einstein-Minkowski space-time is closed and the rest of the space-time is open relative to that

---

[29] S. Savitt, "Time in the special theory of relativity", in C. Callender, F. Arntzenius (eds.) *The Oxford Handbook of Philosophy of Time* (2011), here pp. 563–567, emphases in the original.

[30] See his "Special relativity and the future: A defense of the point present" (2008).

[31] Ibid., p. 2.

point."[32] Harrington's investigations lead him, like Savitt, to emphasize the priority of proper time measured locally along a world line. Succinctly put: "proper time *just is time*."[33]

I concur with the proposals outlined above and believe that they point in the right direction in that they emphasize the relational character of temporal asymmetry, and are based on local events and regions in spacetime, rather than the universe as whole. What is still lacking, however, is an answer to the question of the origin of temporal asymmetry: Why is there a relation of precedence—the relation "before"—between events in the first place? Why does passage from the past to the future occur, and how can we account for the fact—or, at least, the impression—that the future is open and the past fixed? These features of time, as the authors cited above rightly argue, are not in conflict with spacetime physics, but they still remain to be explained.

In addition to the above, recent years have seen contributions to the literature on time and relativity theory emerge which fully embrace spacetime physics and its lack of a global past-present-future distinction, but try to bridge the physics with time's passage as experienced. Michael Silberstein, W. M. Stuckey and Timothy McDevitt propose that this should be done through overcoming traditional divides between materialism and dualism concerning the problem of mind and matter, and instead embracing a "neutral monism".[34] In this way, the authors hope to overcome one-sidedly subjectivist or objectivist accounts of the relation between the subject and the outside world. Instead, something which they call "Presence", a concept taken to be fundamental and not capable of further definition, serves as the desired bridge. The upshot is that, if said dichotomies are overcome, that between time as described by physics versus time as experienced is so likewise.[35] As for the direction of time, the authors contend that, since time lacks an intrinsic, objective direction, no explanation needs to be found for it.[36]

---

[32] Ibid., p. 13.

[33] Ibid., p. 12.

[34] *Beyond the Dynamical Universe: Unifying Block Universe Physics and Time as Experienced* (2018), ch. 8, *passim*.

[35] Ibid., pp. 362–363.

[36] Ibid., p. 367.

Craig Callender, whose work on time and quantum mechanics we already encountered in Sect. 4.4, provides an in-depth survey of experimental work on the psychology of temporal asymmetry, in particular on how our attitudes and emotions, both positive and negative, vary depending on whether an event is past or future. He combines these psychological considerations with an evolutionary account, whereby living beings which respond differently to past and future events have a survival advantage and are selected for. Callender is hopeful that an explanation for time's experienced asymmetric structure may be attained in this way, although he concedes that work remains to be done to that end.[37]

I am sceptical of the approach proposed by Silberstein, Stuckey and McDevitt. While there is nothing to be said against an account of the relationship between mind and matter which goes beyond materialist reductionism and substance dualism, the question remains: how does this yield time as experienced? How do we get from their undifferentiated "Presence", designed to mediate between the subjective and the objective, to the structure of time with its ordering, metric properties, and characteristic asymmetries? A big, unexplained gap remains. It also seems rather facile to claim that time has no direction to be explained since, while there is indeed no preferred direction in the spacetime manifold considered as whole, individual world lines do exhibit direction. Callender, on the other hand, in my view rightly attempts to bridge experimental psychological knowledge on the rich structure of time as experienced with the philosophy of its modal structure, an intersection of research areas which may well prove fruitful. I would object, however, that different affective attitudes to future and past presuppose that distinction, so that it is hard to see how they could explain it. Similarly, evolution can only select for living beings with such different affective attitudes—as, indeed, it may well have done—if that which those attitudes refer to is already given. Thus, while a deeper understanding of the experience of time may well be attained through such interdisciplinary research, I doubt that it will be able to yield an explanation of the arrow of time itself.

[37] *What Makes Time Special?* (2017), ch. 12.

In sum, then, a satisfactory theory able to account for the arrow of time is lacking in the present state of research. In Chap. 9, I will set out a derivation of the relation "before", and from there on, of the difference between the fixed past and the open future, all in purely local terms, in order to fill this gap. But before doing so, some further building blocks will need to be assembled in the following chapters.

6

Against Substantival Passage

In the last chapter, we have seen that the assumption of a worldwide, substantival passage leads to many philosophical problems, conflicts with physics, and to a doubling of time. As Gödel argued, the notion of a present hypersurface passing through spacetime, conceived as the locus where the possible becomes actual, cannot be upheld if there is no global simultaneity. But, unlike Gödel held, this does not lead to an unchanging world. Change does not need an underlying and pre-existing time. Rather, change is what in a certain sense constitutes time, so that time depends on change, as will be spelt out in more detail in the following chapters. Interestingly, that the assumption of a substantival passage of time leads to a doubling of time—or rather, an indefinite multiplication of times—was already revealed in McTaggart's paradox, independently of the problem of lack of global simultaneity coming from relativity theory. In this chapter, I will argue that the notion of a substantival passage leads, furthermore, to quite unnecessary difficulties in the philosophy of causation, and that still more problems become evident when the relation of such a passage to the laws of nature are considered.

D. Saudek, *Change, the Arrow of Time, and Divine Eternity in Light of Relativity Theory*, Palgrave Frontiers in Philosophy of Religion,
https://doi.org/10.1007/978-3-030-38411-1_6

6.1 Passage of Time and Causation

First, the passage of time is usually associated with the conversion of the possible to the actual occurring in the present. But what is the relationship between these two notions? Is the moving "now" itself responsible for making the possible actual? If so, then it is causally active, operating as a kind of selection principle, choosing which possible states of affairs become actualized. But presumably, causation already does this job. Events and states of affairs are brought about by a great variety of causal activities of entities in spacetime, from the deterministic workings of the solar system to the choices of rational agents. What does the passage of time add to this, and why should a kind of temporal causation be introduced over and above these causal activities? A causally active substantival passage leads effectively to a double determination of events: one by ordinary causality, and the other by the assumed temporal selection principle.

Of course, this way of conceiving the moving "now" may seem like a caricature. Its proponents will counter that it is not to be understood as an active principle, but that all selection of actual states of affairs from possible ones takes place *at* it. Such a move, however, only shifts the problem and leads to a regress: if the moving "now" is not itself causally active, then how does it relate to the actualization of possibilities occurring at it? If it in no way influences them, then it has no explanatory role. If it does influence them, then it is once again a causally active principle over and above ordinary causation, and we incur the problem of double determination. One might avoid this dilemma by viewing the moving "now" as something which in a sense triggers the actualization of possibilities, without however influencing the outcome. But then again, such an entity is unnecessary: causation is at work anyway and doesn't need the moving "now" in order to prompt it.

An example which illustrates this dilemma is the causality exercised by rational, free agents in spacetime. An important motive for introducing an absolute hypersurface defining the world's "now", and separating its past from its future, is precisely to ward off the threat of a block universe and to thereby save the possibility of agent causation towards the open future. But for the reason just given, this move in fact backfires: if the worldwide "now" is invested with any causal power and accounts for the actualization of possibilities, it conflicts with agent causation rather than making room for it. Supposing that agent causation exists, it is agents, not time, who make possibilities actual by their choices and actions. If, on the other hand, the moving "now" has no such power, we might as well do without it.

A substantival passage of time therefore has no explanatory role and can be discarded. In other words, there is no temporal determinism over and above causation. What substantival passage is supposed to explain, namely time's one-way street character, the fixity of the past and the openness of the future, can, in my view, be accounted for without it, as I will argue in the following chapters.

6.2 Passage of Time and the Laws of Nature

Arguments similar to the above also reveal limits to the enterprise of accounting for passage of time in terms of physical law. According to the physicist Heinz-Dieter Zeh,

> If physics is to justify the hypothesis that its laws control everything that happens in nature, it should be able to explain (or consistently describe) this fundamental asymmetry which defines what may be called a *direction in time* or even … a direction *of* time.[1]

However, a decisive feature of temporal passage cannot, arguably, in principle be accounted for in terms of physical law: the asymmetry between the open future and the fixed past. For if we suppose that this feature was explicable by physical law, we get similar difficulties to those encountered in the previous section: if the laws, given certain initial conditions, determine which possibilities are actualized, then it is no longer true that the future allows several possibilities, and the initial assumption that the future is open is false. It has also been suggested that the laws of

[1] H. D. Zeh, *The Physical Basis of the Direction of Time*, 3rd edn (1999), p. 1. Emphases in the original.

nature do not directly determine which possibilities become actual, but only determine where in spacetime this occurs.[2] But this leads to the existence of a privileged present hypersurface, which is not only in conflict with relativity, but which also, by the argument in the last section, is causally inert, and can therefore be dispensed with altogether.

The first of these options leads furthermore to the difficulty of whether and how a deterministic universe could evolve in time at all. Suppose, for example, that we live in a universe as envisaged by David Hume,[3] where all events, including the acts of rational beings, are accounted for by necessary causation and are therefore fully determined from the initial state of the universe and the laws of nature. Then, each event occurring in the history of the universe is implied by the laws and initial conditions, or in other words, the latter are sufficient for bringing about each event. But then, why are many events not instantiated until a long time after the beginning of the universe? Conversely, if we live in an "open universe" as argued for by Karl Popper,[4] it is not true that all events are determined by laws and initial conditions. Then, some necessary conditions for a particular event will arise during the development of the universe, rather than at its beginning. The fact that a particular event does not occur at the beginning can then be accounted for by the lack of some necessary conditions for it. However, this latter option means that new causal powers come into being during the history of the universe, apparently unexplainedly. We thus get a dilemma similar to that which philosophers of nature already encountered in antiquity, as has been succinctly described by Aristotle:

> The first of those who studied science … say that none of the things that are either comes to be or passes out of existence, because what comes to be must do so either from what is or from what is not, both of which are impossible. For what is cannot come to be (because it *is* already), and from

[2] Cf. Zimmerman, "Presentism and the space-time manifold" (2013), pp. 89–90. According to Zimmerman, the privileged foliation defining the passage of time in the manifold is not itself causally efficacious (p. 77).

[3] *Enquiry Concerning Human Understanding*, section VIII, I–II.

[4] *The Open Universe: An Argument for Indeterminism* (1988).

what is not nothing could have come to be (because something must be present as a substratum).[5]

The physics of time asymmetry and of the unidirectionality of time is a complex and contentious field into which several more areas of physics enter. This book has so far focused on the contributions to this field from special and general relativity theory, and has touched on that from quantum mechanics, but classical mechanics, statistical physics and thermodynamics, the physics of the emission and absorption of radiation, and other more specialized areas of research are also highly relevant to the field. We will not explore the contributions from these latter in any detail.[6] Generally speaking, most known laws of physics are time-symmetrical in character. They do not distinguish between a forward and a backward direction, so that the evolution of any system as a function of t is just as viable, as far as the dynamic laws of physics are concerned, as its evolution as a function of $-t$.

The time symmetry of dynamic laws is not to be confused with a time symmetry in the state of a system.[7] The evolution of states of a closed system obeys the second law of thermodynamics, whereby the entropy of such systems never decreases ($dS/dt \geq 0$), so that earlier states are distinguishable from later ones. The second law of thermodynamics has therefore often been invoked as the main contender for a physical basis of the direction of time, for example, by Peter W. Atkins and by Ludwig Boltzmann himself. However, as Barry Dainton points out, rise in entropy does not appear to be able to explain the other asymmetries connected with the passage of time, that is, those involving causation, knowledge and explanation. Moreover, the passage of time is not affected by growth or decrease in entropy so that, for example, clocks are not observed to go backward in regions where entropy decreases. Increase in entropy can

[5] *Physica*, 191.a, 23–32.

[6] For systematic overviews, see especially H. D. Zeh (1999); P. C. W. Davies, *The Physics of Time Asymmetry* (1977); P. Kroes, *Time: Its Structure and Role in Physical Theories* (1985).

[7] Cf. J. Baez, review of *The Physical Basis of the Direction of Time* by H. D. Zeh (1993).

therefore be viewed as a symptom of the passage of time (in most cases), but not as its cause.[8]

Indeed, as D. H. Mellor argues, "irreversibility has nothing to do with what gives time its direction ... the direction of time does not depend on the existence of irreversible processes."[9] This is clear from the consideration of systems undergoing irreversible development which are interfered with. For example, suppose that we insert an atom of a metal into a gas. In such a case, there is an earlier stage of the gas, namely without the metal, and a later one with it. It does not matter, as far as the direction of time is concerned, whether the gas evolves from a low-entropy to a high-entropy state or, by the almost impossible, vice versa.

To sum up this chapter, many considerations from both philosophy and physics speak against the notion of a passage of time as an entity in its own right and invested with its own causal power. This, of course, agrees well with the arguments from relativity theory against the existence of any privileged or frame-independent temporal ordering, and therefore also against the possibility of assigning A-theoretical predicates to portions of spacetime in an observer-independent way. The fact that theoretical physics appears to be "blind" to temporal passage is less surprising when this notion, at least as construed in substantivalist terms, is subjected to philosophical scrutiny and found to be superfluous. By this, I do not mean that the passage of time as we experience it is illusory. Rather, it needs to be accounted for in relational terms.

[8] B. Dainton, *Time and Space*, 2nd edn (2010), pp. 47–50; *q. v.* for references to Atkins and Boltzmann. See also C. Callender, *What Makes Time Special?* (2017), p. 244, footnote 4, for an overview of references from the research tradition linking time irreversibility to thermodynamics.

[9] D. H. Mellor, *Real Time II* (1998), p. 120. Mellor uses clocks going clockwise and anti-clockwise, whose hands are being bent, as an example of irreversible processes being interfered with.

7

Understanding How Change Relates to Time

What has been developed so far motivates a theory of time which understands it in terms of change, and therefore of causation, rather than as an entity existing independently over and above change. I will call such a theory "causal" because, according to it, causal interactions account for time. This, however, does not mean that causal asymmetry is the foundation for temporal asymmetry, a point discussed in more detail in Sect. 10.2. Some reasons for the move to a causal theory come from philosophical considerations on relativity theory: In Chap. 3, we saw that the combination of a pre-relativistic and absolute passage of time with its frame-independent distinction between past, present, and future raises unnecessary problems concerning causation and temporal determinism. This point was reinforced in Chap. 5, where it was found that the introduction of an absolute time in the context of spacetime physics leads to a rather profligate multiplication of entities and to the postulation of implausible physics. Relativity theory provides yet a further hint in favour of a causal theory of time because, as Milič Čapek emphasizes, "The succession of causally related events is preserved in *all* frames of reference",

D. Saudek, *Change, the Arrow of Time, and Divine Eternity in Light of Relativity Theory*, Palgrave Frontiers in Philosophy of Religion,
https://doi.org/10.1007/978-3-030-38411-1_7

whereas "*the temporal order of all causally unrelated events remains undetermined*".[1]

Further philosophical problems arising from the notion of a substantival time existing over and above change became evident in the previous chapter, where we saw that the assumption of such an entity leads to yet further difficulties in the philosophy of causation and the laws of nature. And finally, a substantival passage entails that events possess incompatible properties, a problem which can only be escaped by an infinite regress, as evidenced by McTaggart's argument.

Conversely, if time depends on change, and therefore on causal interactions, then we do not need an absolute time, so that the aforementioned difficulties in philosophy and physics are avoided. However, to account for the familiar features of the passage of time on the basis of a causal theory is a difficult task, and many objections must be met. A viable theory of time needs to account for its asymmetric character in the context of laws of nature which do not single out a preferred time direction, that is, for its observed unidirectionality, by which we cannot travel back and forth between points in time as we can do between points in space, and for the difference between the fixed and unchangeable past and the open future which can be influenced. Above all, a causal theory of time must not circularly introduce a pre-existing time in order to achieve its explanatory task. In the following chapters, I argue that such a theory is up to these challenges.

7.1 Help from the Ancients: Aristotle, Suárez, and Augustine

Before presenting the case that a causal theory of time is able to save the phenomena associated with time, it will be useful to consider some of the work on such a theory undertaken in the Aristotelian tradition. We will therefore first look at Aristotle's understanding of time, and then into the detailed elaboration of the Aristotelian theory developed by the Spanish

[1] M. Čapek, "Time in relativity theory: Arguments for a philosophy of becoming", in J. T. Fraser (ed.) *The Voices of Time* (1981), here pp. 443–444. Emphases in the original.

Jesuit Francisco Suárez in his *Disputationes Metaphysicae*. Then, we will discuss some aspects of Augustine's thought on time, an approach which is quite different from an Aristotelian one, but which furnishes some valuable insight on temporal measurement and the distinction between past, present, and future.

As is well known, Aristotle defined time as "number of motion in respect of 'before' and 'after'".[2] In Aristotle's work, the term κίνησις, here translated as "motion", refers to much more than motion through space (ἡ κατὰ τόπον κίνησις), but also includes increase and decrease (αὔξησις καὶ φθίσις) as well as qualitative change (ἀλλοίωσις). The term μεταβολή refers to change in general, which in addition to the three types of κίνησις comprises generation and corruption (γένεσις καὶ φθορά).[3]

Aristotle's definition is based on considerations of the relationship between time and change or motion. Time cannot simply be equated with change or motion, for two reasons: first, "the change (μεταβολή) or movement (κίνησις) of each thing is only *in* the thing which changes or *where* the thing itself which moves or changes may chance to be. But time is present equally everywhere and in all things." And second, "change is always faster or slower, whereas time is not: for 'fast' and 'slow' are defined by time." On the other hand, Aristotle argues that time is not independent of motion either, because absent any change, no time seems to pass. This is the case in unchanging states of consciousness, or when people undergoing a period of unconsciousness are not aware that any time has elapsed, but "connect the earlier 'now' with the later and make them one". Aristotle concludes that "time is either movement or something that belongs to movement. Since then it is not movement, it must be the other." The mind is aware of the passage of time by noticing movement and by "marking it by 'before' and 'after'; and it is only when we have perceived 'before' and 'after' in motion that we say that time has elapsed". This act of "marking" is described as a pronouncing of "nows" by the mind. Since change "admits of enumeration" and is continuous, its magnitude is likewise continuous. As Aristotle explains, "we discriminate the

---

[2] *Physica*, 219b, 1–2.

[3] K. Prantl, *Aristoteles' Werke*, vol. I (1978), p. 475.

more or the less by number, but more or less movement by time. Time then is a kind of number."[4]

Thus, in Aristotle's view, time is based on change occurring mind-independently in the world but, being a "number", it also depends on an act of the mind. "Before" and "after" likewise have their foundation in motion itself, but are not identical with it,[5] rather depending on the mental activity of "marking" and quantification.

Suárez follows Aristotle's theory of time while also developing it further. His thought on time is based on considerations concerning duration. This latter, as he argues, is nothing other than existence, the two being distinguished only by an act of reason, but not in reality. For while the concept of "duration" has a different meaning from that of "existence", both are inseparable in reality, since a thing (*res*) endures if and only if it exists.[6] Suárez notes that we find in things a "real successive duration", and this presupposes an entity whose duration it is. He therefore distinguishes between the duration of a "successive thing" and durations which "have their own permanence in the same being". Since "duration accompanies being, or rather is in reality identical with it", only a "successive thing" can serve as a *fundamentum in re* for the real successive duration observed in nature.[7] Suárez argues that "in reality there is no successive being which could underlie successive duration, except motion; therefore this duration cannot be anything other than motion … for nothing is by itself in continuous succession, except insofar as it is in becoming."[8] He therefore

---

[4] *Physica*, 218b, 8 – 219b, 4.

[5] 219a, 18–20.

[6] *Disputationes metaphysicae*, L, sect. I, 8.

[7] "Certum est dari in rebus huiusmodi realem durationem successivam … inter entia dantur quaedam successiva, quae ita existunt ut necessario per aliquam moram in suo esse permaneant … Sicut ergo esse rei successivae non ita durat ut idem omnino permaneat in tota mora suae durationis, sed ita ut pars eius post partem adveniat, ita duratio talis esse non habet permanentiam stabilem, sed fluentem, si tamen nomen *permanentiae* meretur; late tamen loquendo permanet seu durat, quamdiu fluxus eius non cessat aut finitur. Patet consequentia ex generali etiam principio supra posito quod duratio comitatur esse, vel in re potius est idem cum illo … aliae [durationes] habent propriam permanentiam in eodem esse in quo vel ex se sunt, vel ab alio conservantur; haec [duratio] vero potius nunquam permanet in eodem esse secundum idem, sed continue variatur secundum novas partes." *Disputatio* L, sect. VIII, 3.

[8] "Duratio ut sic supponit saltem ordine rationis esse cuius est duratio, et extra illud non reperitur; sed in rerum natura nullum est ens successivum quod supponi possit ad durationem successivam,

repeatedly identifies "real successive duration" with time itself.[9] But since duration is only distinguished from being by an act of reason, and since time is the duration of motion, Suárez concludes that time "in reality is not distinguished from motion, but only by reason".[10]

Suárez then considers the objection that the same motion can take place in more or less time, so that time cannot be identified with motion. He provides two quite elaborate answers, of which we will only consider the first: in it, Suárez distinguishes "real duration" from the duration of a change due to a comparison to an "extrinsic time". For example, a revolution of the sky around the earth could take place faster or more slowly, but "in the same motion, so to speak, of one revolution of the sky, the real duration is the same, for the real being of such motion is the same ... although that entire duration, by comparison to an extrinsic time ... could last longer or shorter according to the faster or slower passing by of that motion".[11]

Thus, like Aristotle, Suárez affirms the existence of true change, becoming (*fieri*), and succession in the world, whereas time in addition to this requires an act of reason. Arguably, what would later become known as Leibniz's principle of the identity of indiscernibles is also central to Suárez's thought on time: both the duration of unchanging things in time and the observed different rates of change are phenomena which can only be ascertained by comparison to some external change, and have no meaning independent of such comparison.

Augustine's investigations into the nature of time, found mainly in the 11th book of his *Confessiones*, follow a quite different approach, which appears to contrast with that of Aristotle and Suárez. As we have seen,

---

nisi motus; ergo haec duratio non potest esse nisi motus ... Et ratio est quia nihil est per se in continua successione, nisi quatenus est in fieri." Ibid., sect. VIII, 4.

[9] Sect. VIII, 1 and 6.

[10] "Duratio in re non distinguitur ab illo esse cuius est duratio; sed tempus est duratio motus; ergo in re non distinguitur ab ipso motu, sed tantum ratione." Sect. IX, 1.

[11] "Respondetur primo aliud esse in aliquo motu esse plus vel minus durationis realis, aliud vero quod motus ipse plus vel minus duret, secundum coexistentiam ad aliud tempus extrinsecum ... Nam in eodem motu, verbi gratia, unius circulationis caeli, eadem est realis duratio, quia idem est reale esse talis motus, et consequenter eadem est realis permanentia eius, quamquam tota illa duratio per comparationem ad extrinsecum tempus, vel ad successionem imaginariam, magis vel minus durare posset iuxta velociorem vel tardiorem transitum illius motus." Sect. IX, 4.

Augustine opposes the identification of motion in the natural world with time, for any standard of time taken from regular motion—such as that of the heavenly bodies—appears arbitrary and raises the question: "why then should not rather the motions of all bodies be time?"[12] Time moves on even if the regular motion used as a standard ceases. Likewise, the duration of rest of a body can be measured by time, which continues despite that rest. Moreover, motion seems to presuppose time: "I hear that no body moves, except in time." Based on such considerations, Augustine concludes: "time is not the motion of a body."[13] On the other hand, Augustine notes elsewhere that time cannot exist without the motion and change of creatures.[14]

Augustine also considers temporal measurement in detail: longer times are measured by shorter ones, and the quantity of time consumed by a process can only be given through such comparison.[15] Unlike in spatial measurement, the difficulty arises that the object of measurement appears not to exist: the temporal extension of a process can only be measured while it occurs, but the past part of the process is no more, the future part is not yet, and the present has no extension. Augustine's solution to this difficulty is that temporal duration is measured within the mind. More precisely, the impact or effect (*affectio*) which passing things leave on the mind is the object of temporal measurement.[16]

Augustine's thought on the distinction between past, present, and future is rather complex: On the one hand, the fact that it is possible to make true statements about past and future events, that is, to relate past ones and to prophetically foretell future ones, seems to presuppose that the events in question exist. However, statements about the past can also be interpreted as based on the traces they leave in the mind, and those

---

[12] "Cur enim non potius omnium corporum motus sint tempora?" *Confessiones*, XI, 23.

[13] "Corpus nullum nisi in tempore moveri audio … Non est ergo tempus corporis motus." Ibid., 24.

[14] "For where there is no creature, by whose changeable motions times are brought about, there can be absolutely no times … Time … passes due to changeableness." In the original: "Ubi enim nulla creatura est, cuius mutabilibus motibus tempora peragantur, tempora omnino esse non possunt … tempus … mutabilitate transcurrit." *De civitate Dei*, XII, 15,2.

[15] *Confessiones*, XI, 24–26.

[16] Ibid., 27.

about the future as based on present signs which announce future events.[17] Hence, it is not quite clear whether Augustine held that present, past, and future events differ in their ontological status. At any rate, he strongly links these categories to the temporal asymmetry of human action and to the associated mental activities: the past is the object of memory, the present of attention, the future of expectation. Augustine illustrates this with the example of the singer: "before I begin, my expectation is stretched over the whole song, but when I have begun, as much of it as I have plucked over into the past is also stretched in my memory, and the life of this action of mine is extended into my memory due to what I have said, and into expectation due to what I will say."[18] The activity of the mind appears to be crucial for Augustine's understanding of temporal asymmetry: "Present intention transfers the future into the past."[19] However, he leaves open how this action asymmetry relates to the perceived temporal asymmetry of events which are not actively brought about by agents, but which rather occur to them.

Despite the differences between Augustine's approach to time and that of Aristotle and Suárez, important points of agreement between them emerge: Both deny that time can be equated with change itself, and fundamentally for the same reasons. They furthermore agree that there would be no time in the absence of change. Considerations about measurement of temporal processes by comparison with a standard play a crucial role in both. Due to this, according to both approaches, an adequate understanding of time is only possible by taking the activity of the human mind into account.

How does an Aristotelian understanding of time square with relativity theory? As is well known, Einstein departed from an absolute Newtonian time, instead understanding time operationally in terms of natural processes of motion, in particular the speed of light.[20] This is consonant with

---

[17] Ibid., 1–18.

[18] "Dicturus sum canticum, quod novi: antequam incipiam, in totum exspectatio mea tenditur, cum autem coepero, quantum ex illa in praeteritum decerpsero, tenditur et memoria mea, atque distenditur vita huius actionis meae in memoriam propter quod dixi et in exspectationem propter quod dicturus sum." Ibid., 28.

[19] "Praesens intentio futurum in praeteritum traicit." Ibid., 27.

[20] See G. Schwarz, *Raum und Zeit als naturphilosophisches Problem* (1992), pp. 191–192.

the views of Aristotle and Suárez, whereby time is a number assigned to change, an assignment which can only be accomplished through some external standard. It arguably also squares well with Augustine's emphasis of temporal measurement as dependent on mental activity, although the consequences of these considerations for the ontology of time are less clear in Augustine than in Aristotle or Suárez. The idea that "time is what the clock measures" is corroborated by the empirical confirmations of both the special theory of relativity (SR) and the general theory of relativity (GR), whereas an assumed absolute time, flowing equably and independently of the relative motions of bodies or of the content of the universe, appears not to be needed. As Wolfhart Pannenberg observed, "modern physics ... follows the Aristotelian tradition of understanding time ... in that it determines time from the procedure of temporal measurement and, in doing so, uses a unit of motion which serves as a measure—the speed of light in relativity theory—as a basis."[21] Of course, findings of relativity theory such as relativity of simultaneity, and time dilation effects in SR and GR, were beyond the purview of the philosophers discussed in this section. But their thought on the relationship between time, change, and measurement is arguably consonant with what emerges from spacetime physics.

7.2 "Spanning" Time: The Leibniz Principle and its Converse

Since our goal is a theory of time in terms of change, we next need to take a closer look at this latter notion. For now, we can explicate the term "change" as the acquisition or loss of a property *p* by an entity *A*. Note that this explication has two elements: (1) For entity *A*, there are two states, that is, *A-with-p* and *A-without-p*. This presupposes that there is sufficient continuity between *A-with-p* and *A-without-p* so that we can speak of both states of *A* as belonging to the same object. Otherwise, *A*

[21] "[Die] moderne Physik ... indem [sie] die Zeit vom Verfahren der Zeitmessung her bestimmt und dabei eine als Maß dienende Bewegungseinheit—in der Relativitätstheorie die Lichtgeschwindigkeit—zugrunde legt, folgt sie der aristotelischen Tradition der Zeitauffassung." *Systematische Theologie*, vol. II (1991), p. 114.

ceases to exist and is destroyed rather than changed.[22] (2) One of the two states of *A* is before the other. The first element of the above explication, as I will argue in this section, yields important insight into the relationship between time and change. But the stated goal also requires a reductive explanation of the second element, that is, the before-order between states of entities. This is provided in Chap. 9, and, based on it, a precise definition—rather than an explication—of change. For now, however, the notion of "before" will be presupposed as known from everyday experience.

A brief interlude on identity criteria before we move on: Leibniz's *principium identitatis indiscernibilium* (PII) states that indiscernibility is a sufficient criterion for identity: if entities *A* and *B* are indiscernible, then they are identical. PII does not imply that the converse principle holds also, that is, it is not necessarily the case that, if entities are identical, they are also indiscernible—the prototypical counterexample being different states of the same object, such as *A-without-p* and *A-with-p*.

We could of course be purists about identity, argue that *A-without-p* is not perfectly identical to *A-with-p*, and therefore demand that the converse of Leibniz's principle, which we could call the *principium indiscernibilitatis eorundem* (PIE), hold too. On this picture, indiscernibility is not only sufficient, but also necessary for identity, so that it is not possible for an entity *A* to acquire or lose a property *p*, since *A* would then *ipso facto* cease to exist and a new object would come to exist in its place.

Neither in ordinary language nor in a scientific context is PIE demanded. The fact that we are not usually purists about identity enables us to give accounts of the evolution or development of objects and systems thereof. For example, if we were purists about identity, we could not describe the evolution of the earth over the last 4.6 billion years, since any change in the earth, by the PIE, would mean that it ceases to exist. On a non-purist view, on the other hand, objects can be viewed as acquiring or losing properties during their existence, from their coming into being to their destruction.

It is of course contradictory to assign *p* and *not-p* to the same object *A simpliciter*. But there is nothing contradictory about the notion of

[22] Cf. E. Runggaldier, C. Kanzian, *Grundprobleme der analytischen Ontologie* (1998), pp. 154–167.

different states of an object, provided that these states be roughly the same, that is, that they be characterized by sufficient continuity so as to be viewed as different states of the same object. This is possible only because we are not purists about identity and do not demand PIE. Again, in both everyday and scientific language, contradiction is avoided by saying that *p* and *not-p* are not assigned to the same object *A simpliciter*, but in different respects. Time functions as such a respect, enabling us to assert "*A* does not have *p* at *t*" and "*A* has *p* at *t′*". Loosely speaking, we could also say that the fact that roughly—but not perfectly—identical objects exist with contradictory properties is what "spans" time.

Crucially, by this I do not mean that we introduce time in order to project a non-contradictory structure onto an underlying reality which could, by its own nature, allow for contradictory statements about it. Rather, we allow a sufficient roughness in our identity criteria so as to view states of an object with contradictory properties as nevertheless belonging to the same object. To do so is possible only if there is a respect such that no contradiction results from doing so. Nor do I mean to say that viewing different states of *A* as belonging to the same object *A* is merely an artefact of thought, so that we would arbitrarily group entities-at-times together and declare that they together constitute the history of object *A*. For since *A* endures, which following Suárez is nothing other than to exist, it is in continuity with itself, even as it undergoes change, a continuity which does not hold between a state of *A* and a state of a similar object *B*. In other words, because endurance is simply existence, we do not need any further explanation for the continuity of an object with itself through time, for example, the explanation that a later state of *A* is the causal descendant of its earlier state.[23]

Objects which retain their identity even as they acquire or lose properties are known in the literature as *endurers*. When an endurer suffers change, that is, when it acquires or loses a property, such change can always be understood causally. That is, it can be attributed to and

[23] On this point, cf. also D. H. Mellor's discussion of the causation of stasis in *Real Time II* (1998), pp. 100–101. According to Mellor, a thing *a* remaining unchanged between two times *t* and *t′* is only an instance of "factual causation": "*a* is *F* at *t′* because it is *F* at *t*." But since there is no change of *a* between these two times, "in stasis, if not in change, there is no particular causation".

explained in terms of something which affects the endurer—another object, a force, or an event—and which can be viewed as the cause of the change. This is true even if we allow the possibility that the cause of change is not itself susceptible to causal explanation. For example, macroscopic objects can in principle be affected by events occurring at the quantum level, which according to current theory cannot be explained and may not even be explicable in principle.[24] In such cases, the observed change in the endurer can be attributed to random change at the quantum level as its cause. A theory which explains time in terms of change of endurers is therefore a causal theory also in the context of quantum mechanics.

A quite different picture emerges within the quantum level itself: here, we cannot speak of endurers as suffering change in the same way as this is possible at more macroscopic levels. The reason is that the identity of microscopic particles is not well defined over time. The motion of a particle through space is governed by its wave function $\Psi(\mathbf{r},t)$, whose square modulus gives the probability density for finding the particle at a given place at a given time. This intrinsically probabilistic feature of quantum mechanics precludes the possibility of identifying with certainty a particle measured to be at a certain location at a given time with a particle sitting at the same location at a slightly later time. Hence, while it is possible for particles to acquire or lose properties—for example, electrons can be made to change their spin—we cannot speak of them as endurers in the same sense as we did for object *A* above.[25] If time is to be understood in terms of change suffered by endurers, it cannot therefore be a feature intrinsic to the world at the quantum mechanical level. It is possible, however, to describe quantum mechanical events in terms of time

[24] For example, according to H. Margenau, "Single events of great magnitude cannot be said to have the picturesque single-event causes which classical mechanics envisaged: even though I know the exact state of a neutron (having a sharp energy) that is capable of setting off an atomic bomb, and know precisely the location of the block of plutonium that is ready to be exploded, I cannot predict whether the disaster will occur. The fate of the globe, as a single event, may hide itself within atomic uncertainty." *The Nature of Physical Reality* (1950), p. 420. As in Chap. 3, I assume the standard Copenhagen interpretation of quantum mechanics, according to which all accessible information about the state of a quantum mechanical system is contained in its wave function Ψ.

[25] Cf. the quote from E. J. Lowe on the identity conditions of electrons in Sect. 3.5.

as measured by more macroscopic clocks, which are themselves endurers, and therefore to construct the wave function Ψ of a quantum mechanical object as a function of time.[26]

Note that the explication of change advanced in this section is quite different from another understanding of change sometimes found in the literature, namely as an alteration of the event content or fact content of the world with the lapse of time or, as Dorato put it, "an objective coming into existence of facts or events".[27] This view seems to me disadvantageous, since on it, we have simply a coming into being of ephemeral brute facts about what is occurring or existing, rather like a succession of motion pictures. Since events and facts about states of affairs depend for their existence on objects and causal interactions between them, the view of change which I advocate is about what happens to objects as a consequence of such interactions.

It might also be objected that my explication of change as the acquisition or loss of a property by an entity is too narrow: for example, when several bodies are in relative motion and change their relative position, we have change without any one of the bodies acquiring or losing a property. The answer is that what is regarded as an endurer is to some extent arbitrary, so that in this case, we can also see the system composed of several endurers as an endurer in its own right, which acquires a property—for example, a different geometry—in virtue of the change in the relative location of its parts, but which remains the same system despite this

[26] In the words of E. J. Zimmerman: "'Time' which appears in the equations [of quantum mechanics] is not a quantum mechanical observable, which would be represented in the theory by an operator, but rather a parameter external to the microscopic system … Space and time are not concepts which can be meaningfully applied to single microscopic systems. Such systems are to be described by abstract concepts (charge, spin, mass, strangeness, quantum numbers) which make no reference to space and time. These microscopic systems interact in ways that must also be described abstractly, that is without reference to space and time. When a vast number of such microscopic systems so interact, the simplest and most fundamental result is the creation of a space-time framework which gives validity to the classical notions of space and time, but on the macroscopic level only." "Time and quantum theory", in J. T. Fraser (ed.) *The Voices of Time* (1981), here pp. 492 and 496.

[27] M. Dorato, "On becoming, cosmic time and rotating universes", in C. Callender (ed.) *Time, Reality and Experience* (2002), here p. 258; *q. v.* for a detailed discussion of this view of change in Gödel's work.

change. In this way, it is possible to assert, for example, that the solar system has changed, without thereby being committed to asserting that any planet in it has changed intrinsically.

The explication of change given at the beginning of this section also allows for quantitative change (αὔξησις and φθίσις, in Aristotelian language), insofar as quantity is understood as a property. However, it doesn't include absolute coming into being or destruction (γένεσις and φθορά), or the possibility of a thing changing into another thing, and losing its identity as it does so. In short, this explication covers what Aristotle called κίνησις, rather than the wider term μεταβολή, with however the important difference that motion through space (κίνησις κατὰ τόπον) is not viewed as change intrinsic to an object. Indeed, as was unknown to Aristotle, a body's motion through space does not imply that it is being causally affected, since nothing needs to act on a body in order for it to move inertially.

Moreover, excluding absolute becoming and destruction from the notion of change presents no problem for a theory of time which sees it as "spanned" by change in endurers. For while it is true that we judge that time passes when, for example, an object is altogether destroyed, we can still do so only in virtue of, and against the backdrop of, other things which endure. If an object is destroyed, we might say, as we did just above, that the system of which it was a part has undergone change. We therefore still have the situation that something which remains itself becomes different in some respect, and this, as I propose—and as will be fleshed out in more detail in Chap. 9—is what gives rise to time.

7.3 Two Objections to Explaining Time in Terms of Change

Several arguments can be put forward to the effect that change in endurers and causation are unable to account for the passage of time and its characteristics:

The principal and general objection is that change, rather than explaining time, presupposes it, since there can be no change unless there is an underlying time during which change can occur. As Gerald J. Whitrow pointed out, it seems that any attempts to reduce time to change must tacitly invoke temporal concepts, making such attempts circular.[28] This is backed by Immanuel Kant, according to whom phenomena cannot be understood apart from time, although the existence of time even in the absence of any phenomena is conceivable.[29] The general objection can be broken down to particular features of time which a causal theory is challenged to explain. These fall under two headings: duration of processes of change, and temporal asymmetry in them.

1. Concerning duration, two objections must be addressed:

   a. Different tokens of the same type of change can consume more or less time. But according to a causal theory, time is explained in terms of change, so that such tokens should consume the same amount of time. It seems, therefore, that the existence of a substantival time must be assumed in order to account for the varying durations of different tokens of the same type of change.

   b. It seems that without substantival time, we cannot explain why causal processes are temporally extended at all, and why there is a finite time interval between effects and their causes.

---

[28] "The basic difficulty confronting the causal theory is that the very essence of time lies in temporal succession … unless the existence of *successive* states of phenomena is tacitly assumed it is impossible for the theory to yield temporal succession." *The Natural Philosophy of Time* (1963), p. 275.

[29] *Kritik der reinen Vernunft*, 2nd edn (1787), AA 57-8. In the translation of Meiklejohn (1855, p. 28), the relevant passage reads: "Time is a necessary representation, lying at the foundation of all our intuitions. With regard to phaenomena in general, we cannot think away time from them, and represent them to ourselves as out of and unconnected with time, but we can quite well represent to ourselves time void of phaenomena. Time is therefore given *a priori*. In it alone is all reality of phaenomena possible. These may all be annihilated in thought, but time itself, as the universal condition of their possibility, cannot be so annulled." In the original: "Die Zeit ist eine nothwendige Vorstellung, die allen Anschauungen zum Grunde liegt. Man kann in Ansehung der Erscheinungen überhaupt die Zeit selbst nicht aufheben, ob man zwar ganz wohl die Erscheinungen aus der Zeit wegnehmen kann. Die Zeit ist also *a priori* gegeben. In ihr allein ist alle Wirklichkeit der Erscheinungen möglich. Diese können insgesammt wegfallen, aber sie selbst (als die allgemeine Bedingung ihrer Möglichkeit) kann nicht aufgehoben werden."

2. As to temporal asymmetry, there are again two challenges:

a. How can the relation of temporal precedence—that is, "before" and "after"—between events or states of affairs be grounded without appealing to a pre-existent succession of moments of time?

b. Can we account for the past-present-future distinction experienced by agents in spacetime on the basis of a causal theory?

In the following three chapters, I will argue that these two groups of objections can be answered and that therefore substantival time plays no explanatory role and can be dispensed with. When discussing causal processes, I will restrict myself to relatively simple instances of physical causation, leaving aside issues such as mental causation or the causal relationship between parts and wholes, not because these issues are irrelevant, but rather to simplify the investigation of a matter which is difficult enough as it is. Therefore, the electromagnetic interaction, which is at work behind the familiar pushes and pulls, squeezes, breakings, and so on, of everyday physical causation, will play a prominent part in what follows.

8

Duration of Processes of Change According to a Causal Theory

8.1 Faster and Slower Change

To answer objection 1.*a*, let's consider two tokens, C and C', of the same type of change in a given endurer A: in both instances, *A-without-p* becomes *A-with-p*. That C happens faster than C' can then mean either that *A-without-p* lasts for a shorter time in C than in C', or that the actual process of acquisition of p consumes less time in C than in C', or both. In either case, the answer to the objection can be given in terms similar to Suárez's solution of this problem, and by employing Leibniz's *principium identitatis indiscernibilium* (PII): the duration of any stage of A can only be ascertained by comparing it to the changes of other objects, and the same is true of the duration of the process in which A acquires p. If such changes are absent, the notion that a stage of A, or the process in which A acquires p, is of longer or shorter duration has no consequences and does no explanatory work. It can therefore be discarded. To illustrate this point, consider an object which starts fully white and turns fully red. Suppose that we watch this change twice: in one instance, we count two pendulum periods for it to complete, and in another, three pendulum periods. There is only one intrinsic difference between the two

D. Saudek, *Change, the Arrow of Time, and Divine Eternity in Light of Relativity Theory*, Palgrave Frontiers in Philosophy of Religion,
https://doi.org/10.1007/978-3-030-38411-1_8

instances: the number of pendulum swings, whereas the change of the object is not intrinsically different. But given this, the different number of pendulum swings completely accounts for the different durations of both changes. Duration, then, consists essentially in comparison with an external standard change.

This answer can be extended to more complex systems composed of many endurers undergoing change: Suppose that an absolute background time exists, so that the intrinsic durations of stages of objects, as well as those of causal processes, are given in terms of this background time. Consider, furthermore, a system S of many changing objects and another system S' such that the only difference between the two systems is that all intrinsic durations in S' are multiplied by some constant factor relative to those in S. Systems S and S' are indiscernible, so that once again the assumption of an absolute background time, and hence also of intrinsic duration, has no consequences and explains nothing. Rather, changes of objects within a system can only be characterized as longer or shorter relative to other changes.[1]

Also, the above answer can be extended to stasis, as opposed to change, by considering, this time, an object which remains unchanging. Again, suppose we witness the stasis twice, but register two different durations as measured in pendulum periods. Here too, the only difference lies in the different number of pendulum periods accompanying the two stases. It, rather than an intrinsic difference in the stases themselves, accounts for the differing durations measured.

8.2 Why are Causal Processes Extended in Time?

It remains open, however, how the causal processes through which endurers acquire properties can consume finite time at all, unless there is substantival time in which such processes take place. We are faced with the following dilemma.

---

[1] Leibniz employed his PII to argue against the Newtonian idea of an absolute space. As Dainton (*Time and Space*, 2nd edn [2010], pp. 233–234) points out, a "nocturnal doubling" of all lengths in space makes no difference, as long as the underlying space is Euclidean, so that the state resulting from nocturnal doubling is indiscernible from its predecessor.

Either, causation takes time because causes precede their effects: then, there is always a temporal distance, or gap, between a cause and its effect. But if that is so, an underlying time needs to be assumed to account for the duration of the gap. Moreover, the gap itself remains mysterious and confronts us with yet another conundrum: does any causal process occur within it? If so, the gap must itself contain little gaps in order to explain the temporal duration of these processes. If not, the gap must span a certain time, whose existence is then an unexplained brute fact.[2]

Or causes do not precede their effects, but rather are simultaneous with them. But then, how can any causal process be temporally extended? Since, in such processes, each cause is, by assumption, simultaneous to its effect, by transitivity of simultaneity the cause at the beginning of such a process must be simultaneous with the effect at its end. Thus, such processes ought to telescope into a single instant, which is contrary to experience. As David Hume put it, this would amount to "no less than the destruction of that succession of causes, which we observe in the world; and indeed, the utter annihilation of time".[3]

A proponent of a causal theory of time who defends the first option is David Hugh Mellor, according to whom causal order is the basis of temporal order. The fact that we can affect the future but not the past and perceive the past but not the future, according to Mellor, are both due to the fact that causes precede their effects. This, in turn, is due to the speed of light as the limit speed of all transmission of causal influence across space: due to it, "in no reference frame can causation at any distance, however small, be simultaneous." Mellor then discusses apparent cases of simultaneous causation, in particular Kant's example of the ball lying on a cushion: the ball is the cause of the hollow in the cushion, but in this instance, it seems that cause and effect are simultaneous. Mellor argues that the simultaneity in such cases is only apparent. Rather, at any moment when the cause is in place, the effect takes place a brief instant later. This can be seen when the ball is taken off the cushion. The hollow then disappears—assuming that the cushion is elastic—but with a time lapse in between. There is a similar time lapse when the ball is put on the cushion.

[2] Cf. S. Mumford, R. L. Anjum, *Getting Causes from Powers* (2011), pp. 106–107.

[3] D. Hume, *A Treatise of Human Nature*, book I, part III, section II.

In this latter case, the assumption that the hollow appears after the ball is put on it seems at first sight to lead to the wrong prediction that part of the ball would be in the cushion for an instant. But that this is not so can be accounted for due to the imperfect rigidity of the cushion, so that "every point within it starts to move only *after* the surface between it and the ball starts to move".[4]

However, a strong case can also be made in favour of causes being simultaneous with their effects. First, physical bodies always act on their surroundings, as electromagnetic and gravitational fields emanate from them. Such action of a body is simultaneous with its existence. That is, there is no time lapse between a body's being present and its setting up a field in its surroundings, although for any point p at a finite distance d between a body and its surroundings, there is a time lapse d/c between the relevant action of the body, for example, the emission of a photon or graviton by it, and the effect at the point p, that is, the reception of the photon or graviton.

Moreover, as argued in detail in Chap. 5 of *Getting Causes from Powers* by Stephen Mumford and Rani L. Anjum, it seems that cases of causation involving collisions between bodies show that causes and effects are in fact simultaneous. Mumford and Anjum ask us to consider a collision between two billiard balls *a* and *b*: *a* rolls along the green baize, hits the stationary ball *b* and imparts all of its momentum on it. It stops dead, whereafter *b*, which has the same mass as *a*, moves on at *a*'s original velocity. We are tempted to say that *a*'s rolling towards *b* before impact is the cause of *b*'s rolling away from *a* after impact. But Mumford and Anjum tell us that this is wrong:

> Did the cause precede the effect … ? What was the cause? Was it the roll of *a*? But the problem is that while ever *a* is rolling along the metre of green baize, it has not yet met with ball *b*. How, then, can any causation, with respect to ball *b*, have yet occurred? Doesn't the causation occur at that place and time that balls *a* and *b* meet and not before? We could have stopped ball *a* at any point along the one-metre roll it took before impact and the causation would not have occurred. What happens in the path and

[4] D. H. Mellor, *Real Time II* (1998), pp. 105–111.

direction taken by ball *a* is just the story of how it got to the point of impact. It is the impact which is causally efficacious, not how ball *a* got to the point of impact.

Similar reasoning undermines the idea that the effect is later than the cause:

> Ball *b* does indeed roll away after the impact but isn't this the story of what happened after the causation occurred? Doesn't this case of causation complete itself with the impact? The impact is complete, let us say, and at that point so is the causation. A momentum was passed on from *a* to *b*, and this deed is already done once ball *b* moves along its path. We could stop ball *b* at any place along its path but the causal act that involved *a* would have been completed just the same. We cannot at this point stop the fact that causation between *a* and *b* has already occurred.

Nor, according to Mumford and Anjum, can a temporal interval between cause and effect be assumed as part of the collision itself:

> Ball *a* first meets *b* at time t_α, let us say, and they go their separate ways, after some interval, at time t_ω. For as long as they are together, squashing into each other and then springing apart, causation is going on, according to our account. There is a process, which takes time, but this causal process has not begun until t_α and has not ended until t_ω. Cause and effect were together, at this point.[5]

But if causes are simultaneous to their effects, how are extended causal processes possible, as for example when a hangover follows several hours after the ingestion of alcohol? Mumford and Anjum propose that this is possible by viewing causation as "involving a single temporally extended change in which cause and effect occur together" so that "[t]he cause takes time to do its work, but the effect coincides with that time entirely".[6]

A further objection to their model is that several processes can be causally related and yet temporally separated: "Isn't there a problem that one

---

[5] Mumford, Anjum (2011), pp. 108–109.

[6] pp. 120 and 124.

causal process can lead on to another that is later in time? … It might then be concluded that at least some cases of causation must be non-simultaneous." Mumford and Anjum answer that earlier causal processes can enable later ones, though the latter are not necessarily caused by the former. An example is the following series of processes: sugar dissolves in tea; the tea cools; the tea is drunk; the sugar is digested and converted into energy. There is no time gap between these processes, although clearly some of them can overlap in time.[7]

Let's see whether the proposals of Mellor and of Mumford and Anjum can help us to find a way out of the dilemma facing a causal theory of time described at the beginning of this section. Note that Mellor's model escapes the objection that there is a mysterious gap between cause and effect: the question of what constitutes the gap between cause and effect, and what happens during it, has an easy answer: electromagnetic radiation travels a distance *d* and takes time *d/c* to do so. Mellor therefore has no problem conceding that bodies always exercise causal activity, without a time gap before doing so, since the density of spacetime ensures that effects are always later than their causes:

> Take for example a point particle *q* whose electric charge *E* at a spacetime point *st* makes it affect the electrostatic field at all points absolutely later than *st*. Yet how can it do this without affecting the field at *st* where the *Eq*-fact is, i.e. without a simultaneous effect? For would that not imply unmediated action at a distance? No, again because of the density of space and time. This, by stopping any other spacetime point being next to *st*, makes *all* this *Eq*-fact's effects at other points, however close, be at a distance. But then it also lets its effect at any such point be mediated by its effects at intermediate points. So while all the later action here must be at a distance, it can also all be mediated. This is why, in a dense spacetime, the *Eq*-fact at *st* need only affect the electrostatic field at all points later than *st*. It need not affect the field where it itself is, and so need not entail any simultaneous causation.[8]

[7] pp. 124–126.

[8] Mellor (1998), pp. 110–111.

While Mellor's solution nicely eliminates the mystery of an unexplained gap, the problem with his account, as I see it, is that by invoking the temporal duration which it takes for a photon to travel between two points, it needs to tacitly assume an already existing time. Mellor's explicit goal, however, is a reduction of time order to causal order.[9] Furthermore, applying Mumford and Anjum's considerations to electromagnetic interactions, if we model the photon mediating this interaction as a particle travelling through space towards its target, the causation happens only when the photon is in contact with its target. Hence, we once again have an instance of simultaneous causation where no time lapse is involved in the causation itself. What is left open on Mumford and Anjum's model, however, is why causal processes, which only involve simultaneous causation, nevertheless have finite temporal extension.

As a side note, a significant detail about electromagnetic interactions is that the invariant spacetime interval between a photon's emission and absorption is zero. But the spacetime interval between two events is equal to the proper time recorded by a clock whose inertial path connects these events. One is tempted to say that the time for the photon between the two events is zero, or that "no time passes for the photon". This, of course, would not be quite accurate, since a photon is not an endurer which can measure time between two events, or in other words is not the sort of entity which can record a proper time. The zero spacetime interval involved in such interactions does however intimate that they are not intrinsically temporal at the level of fundamental physics. Perhaps, the speed of light is best viewed as a relation of causal accessibility between objects in space, restricting the causal reach between them. As is well known, if the speed of light were not finite, objects would be exposed to vast, perhaps infinite amounts of radiation hitting them from the entire universe,[10] and presumably could not even subsist. But the issue of the significance of the zero spacetime interval in electromagnetic interactions for the philosophy of time would require further investigation beyond the scope of this book.

---

[9] p. 113.

[10] Cf. T. P. Cheng, *Relativity, Gravitation and Cosmology*, 2nd edn (2010), p. 184.

However that may be, my proposal to the problem of how simultaneous causal interactions can take finite time is that change itself is what gives rise to time, since following Aristotle, time is a number or measure assigned to processes. Consider, for example, a white object becoming increasingly red in *n* successive discrete steps. Let's call *k*-red the degree of redness which the object has in the *k*-th phase of this process. A change in the object from being *k*-red to being (*k*+1)-red leads us to judge that some time has passed. On the other hand, the duration of the object remaining in the same state can only be ascertained by reference to some external standard change. Indeed, by the PII, the notion of this duration has no meaning apart from such a reference, as argued in the previous section. As far as the object itself is concerned, for it to endure is the same as for it to exist. But given this, causal processes can be extended in time, since any *k*-th stage can last for a finite time, as measured by some clock, even if we suppose that all transitions between states are instantaneous, so that the changes themselves contribute nothing to the duration of the whole process of reddening. Moreover, it doesn't matter that the causation due to which the object reddens is such that cause and effect are simultaneous. It matters only that it brings about its effect. But then, causal processes can take a nonzero time to be completed, or in other words, they can be extended in time.

This scenario can be generalized to more fine-grained processes by increasing *n* and reducing the redness acquired in each stage accordingly. Since there is no limit in doing this, the above example shows that continuous changes can likewise be extended over finite time. More complicated causal processes can then involve changes in several properties of one endurer or many, which may overlap or be separated in time and can take place gradually or jerkily. They can moreover occur at faster or slower rates, as noted in the reply to objection 1.*a*.

In short, the objection that simultaneous causation would lead to zero-time causal interactions adding up, absurdly, to finite time can be answered. In a philosophy of time of Aristotelian inspiration, all that is needed in order for there to be time is that there be change and quantification thereof. But all we need for change in endurers is causation which affects them, and causation, as in my view has been ably shown by Mumford and Anjum, is simultaneous and involves no time lapse between cause and effect. Simultaneous causation is therefore not in conflict with the familiar fact that causal processes take time.

8.3 Countable Change and the *Aevum*[11]

While there appears to be no reason to suppose the existence of a substantival time independently of processes of change, time cannot be simply reduced to change either: whereas processes of change can happen faster or more slowly, time is not affected thereby. For this reason, Augustine objected to the position of "a learned man" of his time, according to whom time is the motion of the celestial bodies. For if this motion ceased, but other processes, such as the motion of the potter's wheel, continued, time would still move on. Furthermore, the revolutions of celestial bodies around the earth could themselves become faster or slow down, and this change could itself be measured.[12]

Measurement of duration must allow the quantitative comparison of any two processes. The standard of measurement must therefore consist of a process whose duration serves as a unit. Moreover, it must allow to count the number of units of the duration of other processes. As Wittgenstein pointed out, counting of entities is only possible if a sortal is specified whose individual members are to be counted, so that the question "how many objects are there" is nonsensical.[13] Thus, for temporal measurement too, several members of the same type are needed, which must moreover be clearly distinct from another. In short, we need a process which occurs again and again. It must also occur regularly, although this regularity can itself only be ascertained by comparison to other repeated processes and, by the Leibniz principle, has no meaning independent of such comparison. An important example is a recurring change in an enduring physical object which repeatedly acquires and loses a certain property, as is the case in most clocks.

[11] By "countable" I simply mean "which can be counted". I am not referring to the cardinality of infinite sets.

[12] *Confessiones* XI, 23.

[13] *Tractatus logico-philosophicus*, 4.1272.

Electromagnetic waves likewise meet the properties required of a standard for the measurement of time, and indeed the duration of the second is today defined in terms of the number of a precise frequency of microwave radiation which causes a certain type of energy level transition in the Caesium-133 atom.[14]

By contrast, in a system where recurrent processes which can serve as a standard of time measurement are absent, change can still occur, but it becomes impossible to quantitatively compare the durations of processes. This is the case in purely intra-psychic change, for example in dreams. Quantification of the duration of a dream is only possible by reference to clocks in the external physical world surrounding the dreamer.

A similar example is Suárez's thought on the *aevum*, that is, the realm of angels and spirits[15]: there is change in this realm since, for example, angels can undergo mental change individually and can also influence each other. They can also compare the durations of such changes, but in the absence of a regular and recurring standard change, quantitative measurement is not possible within the *aevum*. Hence, Suárez follows the standard scholastic doctrine whereby change in the *aevum* is not measured by time as we know it. However, it is possible to measure the duration of angelic action affecting the material world, since measures of time are available in the latter.[16]

Likewise, physical changes themselves cannot be quantified in the absence of a standard meeting the criteria described above. We could therefore call a system "*aevum*-like" if it undergoes change, but there are no standard changes available as part of the system itself by which to quantify the change.

A particularly interesting example in this context are the systematic changes associated with the expansion of the universe, which were considered in more detail in Chap. 4. Considered in themselves, these changes—for example, the increasing distance between galaxies, or the growth in the dominance of Ω_Λ—are *aevum*-like, so that the notion of an

---

[14] See any physics textbook, for example, H. D. Young, R. A. Freedman, *University Physics*, 13th edn (2012), p. 4.

[15] *Disputationes Metaphysicae*, L, sect. XI.

[16] Ibid., sect. XI, 12–13 and 16.

intrinsic duration of these changes is once again superfluous and can be dispensed with. However, they can be quantified through regular, periodic changes of systems within the universe. The same applies, therefore, to the age of the universe: it can only be measured by familiar years, seconds or other units, but not by itself. A similar argument was made by Suárez, in the context of the very different cosmology of his day, against the notion of an intrinsic duration of the *primum mobile*, the outermost sphere thought to set the heavens in motion:

> At the most, we can repeatedly measure the whole by a part of that motion, for example the day by the hour, the year by the day etc. In this way of measuring there is already a measure in a thing which is in some way distinct from what is being measured; but we cannot measure the whole by itself.[17]

[17] Ibid., sect. X, 4. "ad summum possumus per unam partem illius motus repetitim mensurare totum, ut per horam, diem, et per diem, annum, etc. In quo mensurandi modo iam est mensura in re aliquo modo distincta a mensurato; ipsum tamen totum per seipsum mensurare non possumus." The opposite view is stated by Unger: "The clock that is neither inside nor outside the universe must be the universe itself. What this means is that certain general, quantifiable, and slowly changing features of the universe, such as the recession of the galaxies in all directions, and the color temperature of the photons in the cosmic microwave radiation background, must be used to time some other change, either in the universe as a whole or in part of the universe." L. Smolin, R. M. Unger, *The Singular Universe and the Reality of Time* (2015), p. 234.

9

The Arrow of Time: The Full Derivation

Based on the groundwork laid in this book so far, it is now possible to derive a local, causal and substance-based account of the arrow of time, starting from simple axioms. This account requires no global past-present-future distinction and no privileged relations of simultaneity between spacelike-separated events and is therefore compatible with relativity theory. It will furthermore allow us to address the second group of objections from Sect. 7.3, that is, those concerned with temporal asymmetry.

To this end, we will first derive, in Sect. 9.1, the ordering and metric properties of local time. In Sect. 9.2, an epistemological assumption will be made, which will then allow us, in Sect. 9.3, to account for time's "arrow" by proving that the local past is fixed and the local future open.

This chapter incorporates material published in my article "Beyond A-theory and the block universe: A non-circular derivation of 'before', change, and the local arrow of time", published online in *Kriterion* (July 2019) and previously in the Phil-Sci archive (March 2018) at http://philsci-archive.pitt.edu/14441/.

D. Saudek, *Change, the Arrow of Time, and Divine Eternity in Light of Relativity Theory*, Palgrave Frontiers in Philosophy of Religion,
https://doi.org/10.1007/978-3-030-38411-1_9

9.1 Constructing Local Time

I propose constructing a concept of time in terms of local causal interactions through the following steps: First, a definition of the temporal ordering relation "before" will be developed. I claim that this is possible in a non-circular way. This will allow us to define the notion of change in substances and, in a next step, to introduce a concept of time as a quantification of change through a measuring standard, in a way similar to Aristotle's.

9.1.1 Defining "Before" Non-circularly

The first step, that is, defining "before" non-circularly, is based on two simple metaphysical assumptions:

A.1 Substances exist, and they can exist in different states characterized by non-essential intrinsic properties.

Thus, if *A* is a substance, then there are non-essential intrinsic properties *p* for *A*, that is, intrinsic properties such that *A* is identical to itself whether or not it possesses *p*. For any such properties *p*, *A-with-p* and *A-without-p* will be called "states" of *A*. As a simple example, if *A* is a football, *A* might possess a red spot, a cut in its leather, and such like, so that *A* with or without each of these—or with none of them—is a particular state of *A*. By contrast, a heap of ashes resulting from burning *A* is not a football and thus is not a state of *A*. We can in this way associate any substance with a set of its conceivable states, or alternatively, with a set of its actually existing ones. In a next step, we may assume:

A.2 For any substance *A*, different states of it exist only if it is subject to causal interactions, either with its environment or among its own parts.

On this assumption, at least some (perhaps all) non-essential intrinsic properties of substances are due to causal interactions. We can therefore term such properties "records", and substances, as the bearers of these properties can be thought of as "recorders". Thus, a dent in a football can be viewed as a record of a causal interaction, of which the football is a recorder. I leave open whether substances also have non-essential intrinsic properties which are not records, since this is not essential to the derivation which follows.

Assumptions (A.1) and (A.2) do not mean that either substances or causation are ontologically basic or that they are necessary for fundamental physics.[1] However, in what follows, I will lay out the case that the asymmetric character of time can be explained on the basis of assumptions about substances and causation. The latter are therefore more fundamental than the former, but this does not imply that they are fundamental absolutely. Indeed, a strong case can be made from theoretical physics—I am thinking in particular of the way in which quantum mechanics reduces the substance-like notion of a particle to the more abstract, mathematical notion of a state vector—that they are not.

In order to see how assumptions (A.1) and (A.2) help us to construct a non-circular definition of "before", consider a simple physical object *A* being affected causally by its environment in the course of two separate events. With each event, a record on *A* is associated. Let's symbolize the two records in *A* by a star and by a triangle, and *A* itself by a circle. Notice that, if the event associated with the triangle is before that associated with the star, then a state of *A* exists with the triangle and without the star, and another state exists with both the triangle and the star, but there is no state with the star and without the triangle—as illustrated below. Thus, we find that, in the scenario under consideration, there is an asymmetry *in the existence of states* of *A*: the situation would be symmetric if there were also a state of *A* with the star, but without the triangle.

The crucial point in the above example is that an asymmetry in the existence of states of *A* presupposes no temporal order in order to define it. Instead, we need only consider the set of states of *A*, check through the states in this set—in philosophers' jargon, quantify over these states—and see whether we encounter the aforementioned type of asymmetry. To illustrate, if each state is represented by a photo, we can find the asymmetry

[1] Cf. the discussion in H. Price, R. Corry "A case for causal republicanism?", in H. Price, R. Corry (eds.) *Causation, Physics, and the Constitution of Reality: Russell's Republic Revisited* (2007), pp. 1–10.

by comparing the photos, irrespectively of the order in which the photos are shuffled. I therefore propose exploiting this asymmetry in order to arrive at a non-circular definition of "before".

However, we are not there yet. In the above example, the triangle could be deleted again, so that we would be left with a state of *A* with the star and without the triangle, in which case the asymmetry in the existence of states of *A* breaks down. Nevertheless, the event associated with the triangle is before that associated with the star. This difficulty, however, can be remedied by introducing the notion of an *amended recorder*:

D.1 An amended recorder is a recorder *A* such that each event affecting it produces a record in *A* and does not affect other records in *A*.

Thus, if a recorder is amended, nothing interferes with records in it. Note that this is not equivalent to demanding that records "stay" or "do not become destroyed" in *A*. To do so would be to circularly invoke the concept of temporal precedence. Instead, the property by which the amended recorder is defined is stated without this concept, so that no reference to "earlier" or "later" states of *A* is needed.

However, the amended recorder needs a further improvement in order to attain a non-circular definition of "before": we need to ensure that each event associated with a record in *A* has a *unique* record. For if the records produced in the course of two distinct events are not unique, the later one of two events affecting *A* can produce a property which is already in *A*, in which case the above asymmetry in the existence of states of the recorder *A* once again breaks down. An example of this case could be a camera which records two indiscernible pictures of a landscape. Let's call *r** the record produced twice in *A* in the course of two distinct events. We notice that, even in such cases, there is still an important difference between these events: one of them is associated with the production of *r** in a state *A-with-r** and the other with the production of *r** in a state *A-without-r**. Notice that, once again, this difference is given in terms of qualitatively different states of *A* and does need to invoke temporal precedence between these states in order to be defined. Said difference can, however, only be made out by a recorder which is able to monitor its own states, that is, to tell in which state, characterized by which records, it exists. Such a recorder will therefore be called an “ideal recorder”:

D.2 An ideal recorder is an amended recorder A such that, for any state S, if A exists in S, A can specify all records of S.

Thus, for an ideal recorder, there will be two distinct instances of $r*$, which could be labelled, for example, $r*_a$ and $r*_b$, even if these two records are intrinsically indiscernible. Each event affecting A is therefore associated with a unique record.

It is now possible to define the relation “before” between events. Let x and y be two events associated with the production of two properties, r_x and r_y, in an ideal recorder A. Then,

D.3 x is before y if there is a state of A with r_x, and a state with r_x and r_y, but no state with r_y and without r_x.

We now have an operational definition of “before” which does without temporal concepts, and which therefore does not need to assume an absolute time, existing independently of substances and events, as ontologically primitive. To briefly recapitulate how the definition works: Suppose we are given the set of all actually existing—as opposed to merely conceivable—states of an ideal recorder A. We can bring these states into a before-order by applying (D.3) to the set, that is, by quantifying over it and finding asymmetries in the states of A. We can be sure that the before-order established in this way corresponds to what we mean by “before” in everyday language because A satisfies definition (D.1), so that records in A are not interfered with. Given this, it can easily be seen that the later a state is according to definition (D.3), the more records it has. This, however, is merely a consequence of A’s being an ideal recorder, not the defining characteristic of the before-ordering, which is given in (D.3). Furthermore, because A satisfies definition (D.2), we can always apply (D.3) to it on the basis of any two distinct records, even if they are intrinsically indiscernible. Every state of A therefore has a definite place in the before-order. In the language of mathematicians, this order is total.

There is an alternative, but equivalent, way of defining "before": Consider the two states *A-with-*r_x and *A-with-*r_x*-and-*r_y of A given in definition (D.3). Clearly, the latter is obtained by adding a defining characteristic to the former. The situation is therefore analogous to, for example, the relation between the notion of "Greek citizen" and that of "human", since the former is defined via the latter, but with the addition of an extra property. In this sense, and following Alberto Strumia,[2] we can think of it as "in" the former. "In" in this sense is therefore a relation which holds, as an important example, between a species and its genus: the notion of "thrush" is "in" that of "bird", the notion of "fraction" is "in" that of "number" and so on. Let therefore, in general, one entity x be called "in" another one y if and only if, conversely, the defining characteristics of y are contained in those of x, just as the defining characteristics of "human" are contained in those of "Greek". Moreover, in accordance with the language of standard set theory, let x be called "properly in" y if and only if x is in y, and y is not in x. Then also, if x is properly in y, y is not properly in x, or in other words, the relation between the two is asymmetric. Of course, such an asymmetry is a timeless one, just as, for example, the asymmetric relation between "Greek citizen" and "human" is timeless, since one does not, in the temporal sense, "first" need to be human before being a Greek citizen. Returning now to our ideal recorder A, we can then say: if there are two states of it, call them A' and A'', marked by different records, then A' is "before" A'', if and only if A'' is properly in A'. Correspondingly, A'' is then "after" A', and likewise, a record in A'' is "after" any record in A' if and only if it is not contained in A'. In this way, the asymmetry in the existence of states of the ideal recorder, the defining characteristic of the relation "before" according to definition (D.3), is intimately related to the asymmetric notion of inclusion which occurs in elementary set theory.

Some readers may object to defining "before" on the basis of the concept of an ideal recorder, pointing out that such substances do not in fact exist. My answer to this objection is twofold:

---

[2] *The Problem of Foundations* (2012), Sect. 2.2.2. Strumia's account is based on set theory and work by Thomas Aquinas, as cited ibid.

First, there are substances in the world we know which approximate the ideal recorder very well. Such substances can therefore be used for establishing the relation "before" between events in a highly—albeit not totally—reliable way by applying the criterion in definition (D.3) to them. For example, even simple physical objects like the moon can be used quite well to establish the order of impact events on it, by quantifying timelessly over its states. The human cognitive apparatus protects records in the brain from interaction with the environment and is therefore a particularly good amended recorder. It has the additional ability of being able to monitor its own states, albeit of course imperfectly. This enables us to identify distinct instances of sense impressions which are intrinsically indiscernible, so that we approximate the ideal recorder well. Furthermore, even though in imperfect recorders, some records of events are interfered with, it is still possible to carry out a comparison with records contained in other recorders, in order to infer what an ideal recorder would record. On this basis, a relationship of temporal precedence, as defined in (D.3), can be established. Defining "before" via the ideal recorder therefore matches everyday experience: there is a before-order between events, but since we are not ideal recorders, we cannot always establish it completely. It is important to note, however, that while we as conscious human beings approximate ideal recorders, consciousness is not required in the above derivation. Machines make just as good recording objects insofar as they fit definitions (D.1) and (D.2). The relation "before" is therefore, according to my account, not mind-dependent or subjective, a point which will be discussed in more detail in Sect. 10.3.

Second, the before-order established by an ideal recorder cannot be violated by a non-ideal one. To see this, suppose again that an ideal recorder A has states *A-with-*r_x(without r_y) and *A-with-*r_x*-and-*r_y, but no state *A-with-*r_y(without r_x). If we symbolize r_x using a triangle, and r_y using a star, we then get the same situation as illustrated in Fig. 9.1. Then, by definition (D.3) we know that r_x is before r_y, and x before y. Suppose we replace A by a non-ideal recorder N, setting it at the same location in spacetime and exposing it to the same events. Suppose also that quantification over N's states yields the reverse order between x and y. This means that we have the following: (1) *N-with-*r_y(without r_x), (2) *N-with-*r_x*-and-*r_y,

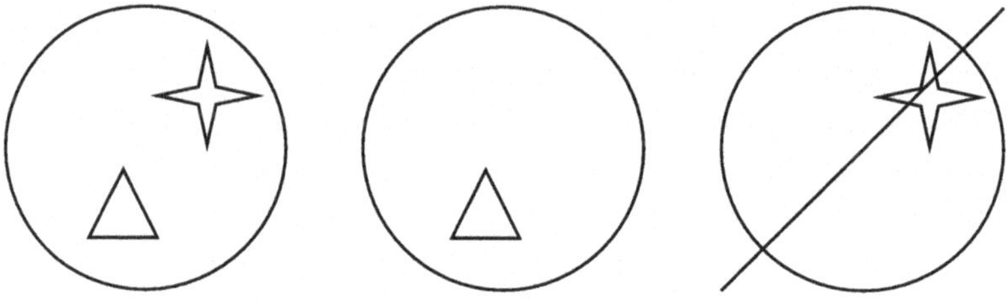

Fig. 9.1 An existence asymmetry in *A*'s states

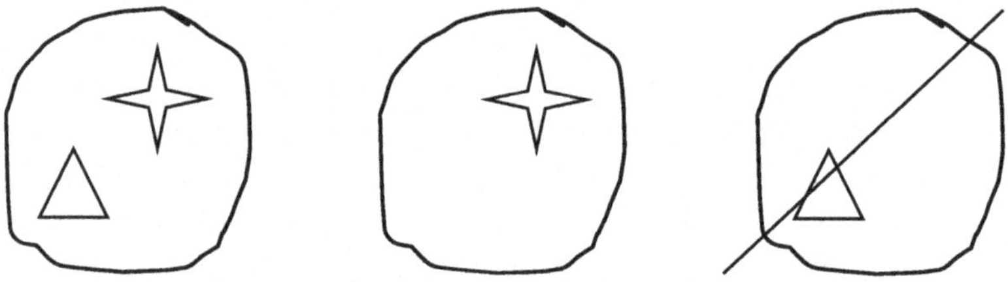

Fig. 9.2 *N*'s states under a hypothetical reversal of the measured before-order

(3) no state *N-with-*r_x(without r_y)—illustrated as follows, letting the squiggly shape symbolize *N* (Fig. 9.2):

But this configuration of states is impossible: for either, *N* records *x*, and therefore has a record r_x, or it doesn't. If it does, there is a state *N-with-*r_x(without r_y), in contradiction to (3). If it doesn't, there cannot be a state *N-with-*r_x*-and-*r_y, in contradiction to (2). Hence, the before-order given by quantification over all states of a non-ideal recorder may fail to yield the complete order established by an ideal one, but this latter order cannot be violated. Therefore, quantification over states of recorders, whether ideal or not, is a reliable criterion of the before-order between local events.

Note also that, even though I defined "before" in terms of quantification over all states of an ideal recorder, this quantification does not need to be carried out explicitly. This is because every event affecting the ideal recorder *A* is associated with a unique record and does not interfere with other records. Hence, given only two states of *A*, such as *A-with-*r_x(without r_y), and *A-with-*r_x*-and-*r_y, we can be certain that there is no state *A-with-*r_y(without r_x). Thus, we do not need an overview over all states of *A* in order to verify whether definition (D.3) is satisfied. This, again, is consonant with everyday experience where, having witnessed two events, we can immediately judge which is before which. To form this judgement, we do not have to await the end of our lives in order to quantify over all our states.

9.1.2 Change

Suppose that a substance *B* is within the causal reach of an ideal, or approximately ideal, recorder *A* and that there are two states of *B*: *B-without-p* and *B-with-p*. These states can affect *A*, for example, by emitting electromagnetic waves, and produce two corresponding records in *A*, which will be called $r_{B\text{-}without\text{-}p}$ and $r_{B\text{-}with\text{-}p}$. This allows amending our preliminary explication of change from Sect. 7.2, and giving, as promised, a reductive definition thereof:

D.4 A substance *B* changes if it acquires or loses a property *p*, where "*B* acquires *p*" if $r_{B\text{-}without\text{-}p}$ is before $r_{B\text{-}with\text{-}p}$ for *A*, and "*B* loses *p*" if $r_{B\text{-}with\text{-}p}$ is before $r_{B\text{-}without\text{-}p}$ for *A*.

Because "before" is defined via the ideal recorder *A*, and because *B* is a substance which is not necessarily an ideal recorder, we cannot define the change of *B* in terms of *B* alone. This poses no problem, since there is never a danger that *A* gets the order between the states of *B* wrong. In other words, whatever be *A*'s location or state of motion, *A* cannot measure a different order from that which would be measured by an ideal recorder whose world line is immediately next to *B*'s. For this latter order is an invariant, as is known from relativity theory, and so it is the same in *A*'s frame of reference. Because of this, the events which are associated with the production of the records $r_{B\text{-}without\text{-}p}$ and $r_{B\text{-}with\text{-}p}$ in *A* must likewise occur in the same order.[3] Hence, if "*B* acquires *p*" or "*B* loses *p*" according

[3] In a special relativistic context, this can be seen from a simple calculation: suppose that *B* changes from green to red, and that, as it does so, it rushes towards *A* at a very high speed v. Let x be the distance between *A* and *B*, according to *A*'s frame of reference, when the process of change starts and when *B* sends off a message "*B* is green". Let $\Delta\tau$ be the proper time for the process of change as measured along *B*'s world line, and let Δt be the time for this process according to *A*'s frame of reference. As is known from special relativity, if $\Delta\tau$ is positive, so is Δt. Let $t = 0$, according to *A*'s

to some ideal recorder, then the same is true according to *any* ideal recorder. It will be shown, in Sect. 11.2, that this is not only a result of relativity theory, but follows from the principle of non-contradiction itself.

9.1.3 Local Time

The above derivation gives us a total order of events through the relation "before", and on this basis, change, but not the possibility of measuring processes of change and of comparing them quantitatively. Such change is therefore *aevum*-like, in the sense in which this term was defined in Sect. 8.3. The next step, therefore, is to give an account of local temporal measurement.

To do so, we use the fact that an ideal recorder can mark two instances of a record of type $r*$ as distinct. It should therefore be possible to construct a local parameter t based on $r*$. I propose doing this in the following way: let A be an ideal recorder or a sufficiently good approximation thereof. Suppose that A is in a state with $r*_a$ and $r*_b$. Then, the set containing $r*_a$ and $r*_b$ can be thought of as a record in its own right, which will be called $r*_2$. A state of A with $r*_2$ can be affected in such a way as to produce a record $r*$ in it. The set containing $r*_2$ and $r*$ can then be called $r*_3$, and so forth. In this way, we get a set of hierarchically nested sets ordered by a relation of inclusion. Readers with a background in mathematics will recognize this procedure: it is analogous to that used in the standard set-theoretical construction of the natural numbers,[4] except that we start, not with the empty set, but with a single instance of $r*$, whose first successor is $r*_2$. Thus, our derivation of the local parameter is not based on a primitive notion of "counting forward", which would be circular. Rather, just as this notion can be reduced to nested sets ordered by a relation of inclusion, the ideal recorder can be equipped with a

---

frame, when the message "B is green" is sent off from B. This message arrives at A at time $t_{green} = x/c$. B is at a distance of $(x - v{\cdot}\Delta t)$, again according to A's frame, when the message "B is red" is sent off. This message arrives at A at $t_{red} = \Delta t + (x - v{\cdot}\Delta t)/c$. Thus $t_{red} - t_{green} = (1 - v/c){\cdot}\Delta t$. This interval can be an arbitrarily small fraction of Δt, but it can never be negative, and so A receives the two signals in the correct order.

[4] See, for example, R. Holmes, *Elementary Set Theory with a Universal Set* (2012), pp. 25–26.

counting mechanism in terms of nested sets based on $r*$, without presupposing a "forward" direction.

We can then identify the index of $r*_i$ as the local parameter t measured by A:

D.5 If A is an ideal recorder, or a sufficiently good approximation thereof, then for all $r*_i$ in A, $i \equiv t$.

As a simple example, suppose that A is a camera and $r*$ is a sunset. A local time can then be defined in terms of nested sets based on records of sunsets in the camera, in the way just outlined. We thus have a natural local parameter, that is, one on the basis of the natural numbers. In a next step, this parameter can be subdivided further by some other repeated change in A, thereby yielding a rational parameter. From it, in turn, a real parameter can be constructed, for example by the standard procedure of defining real numbers in terms of sets of rational numbers.[5] This last step can be done only in thought, since these sets contain infinitely many elements, whereas A, being a physical object, can only contain finitely many records. It remains true, however, that we can refine our original natural local parameter to any standard of precision we please. Let's continue to call this refined parameter t. We can then write states of A as A_t, that is, we can identify them by this local parameter, rather than by the records they contain. By writing A-*with*-r_x simply as Ax, where x is an event and r_x the record in A associated with it, this identification reads:

I.1 $A_t = Ax_1, \ldots, x_n$, for some n.

Thus, the left-hand side of the identification is a state of the ideal recorder A given in terms of the local parameter, whereas the right-hand side gives this state in terms of all of its records, not just those associated with the counting mechanism on which the parameter is based. We do something similar when we equate a state of a recorder—in general, not ideal—in terms of its age with a qualitative description of it in terms of its records (e.g. "At 27 years, Fred has a broken leg, experience in com-

[5] Ibid., pp. 94–96.

puter programming, …"). In (I.1), t is a real number, and n a natural one, since A_t contains records of finitely many events.

Writing states of A as A_t also allows assigning records in A to the local parameter t, by defining:

D.6 If a record r is produced in A_t, then $r \equiv r_t$.

In other words, the index t applied to a state of A carries over to any record produced in that state. If Fred in his state with the time index "27 years" learns, say, a set of Japanese words, then this set, according to (D.6), also gets the time index "27 years". This identification will be needed below when we will show that the past of a recorder is fixed.

On the basis of what has been developed throughout Sect. 9.1 so far, I propose a neo-Aristotelian definition of time:

D.7 Time is a local parameter t established through a recorder A on the basis of different instances of records of the same type $r*$ in A, such that t is used to quantify processes of change in substances which can causally affect A.

Only a substance A sufficiently close to an ideal recorder can perform both tasks of registering change in substances and establishing the local parameter t. Such a substance can quantify changes either in itself or in other substances. In the latter case, the quantification can be carried out in different ways. The simplest way is to use the temporal indices t of records r_t in A which correspond to changes in A's surroundings. Alternatively, change can be quantified from these indices in combination with additional information, in particular about the distance between A and substances which can affect it.[6] In the limit of small distances or a high speed of propagation of causal influence, the results of both methods of quantification will coincide. For example, this happens to a good approximation when we stand on a summit of the Alps and temporally quantify whatever is going on in our surroundings tens or hundreds of

[6] As is done in the Einstein synchronization procedure, described in E. F. Taylor, J. A. Wheeler, *Spacetime Physics*, 2nd edn (1992), pp. 37–39.

kilometres away. This leads to the impression that there is a "world at time *t*", that is, that an entire state of the world corresponds to a locally measured time, as was generally thought before the advent of relativity theory.[7] Note also that the proposed account of time measurement incorporates the insight from Augustine, as described in Sect. 7.1, whereby changes are quantified through the impact (*affectio*) of passing things. However, in Augustine's thought, it is the conscious mind which contains the *affectiones* and which serves as the basis for this quantification, whereas on the present account, a recorder, conscious or not, suffices.

Some readers may be concerned that a repeated process is not sufficient to establish a local time standard, but that the process also needs to occur regularly. However, this objection presupposes that there is an absolute background time with respect to which we can predicate this regularity. But, as argued in Chaps. 6, 7 and 8, we have every reason to discard such an absolute time. Rather, as argued by Francisco Suárez, whose thought on the relative duration of processes we considered in Sect. 7.1, regularity only becomes an issue when several repeated processes come into play, whose rate with respect to each other may vary. If there is only one such process, then there is no way of quantifying the intervals defined by it and of comparing their duration. To illustrate this point, consider an ideal recorder between the two local times k and $k + 1$, where k is a natural number. Suppose that we multiply the duration of this piece of local time relative to a hypothetical absolute time standard by some factor, just as we did in Sect. 8.1. Again, the multiplied duration is indiscernible from the non-multiplied one. Also, there is no quantitative measure which would allow us to tell that the interval from k to $k + 1$ is different in length from any other interval defined by a single unit of local time. By applying the Leibniz principle, whereby what is indiscernible is identical, we see again that the notion of regularity or irregularity with respect to absolute time can be discarded. Therefore, a repeated process alone is sufficient to define a unit of time.[8]

[7] Cf. R. J. Russell, *Time in Eternity* (2012), pp. 263–265.

[8] Cf. the similar observations made by P. Janich in his *Die Protophysik der Zeit*, 1st edn (1980), pp. 180–189.

9.2 Epistemological Realism

In view of deriving the irreversibility of local time in Sect. 9.3, one more premise is needed: I will assume that from records in a recorder A (which need not necessarily be ideal or even amended), conclusions can be drawn about the truth of propositions. As a simple example, from an imprint in the ground, we can judge that an animal has walked past. Of course, such conclusions cannot be drawn directly from the mere existence of records, but rather with the help of some rules of inference taken, for example, from everyday or scientific background knowledge. Let the variable v denote a valid rule of inference, and let V stand for the set containing all such rules. Also, r^A will denote a record in A, and R^A a set containing records r^A. Then, we can assume:

A.3 From the fact that the rules $v \in V$ apply, and from the existence of records $r^A \in R^A$, it follows that some propositions p_R^A are true, the set of which will be called P_R^A.

R^A will contain finitely many records, since it is a subset of the set containing all records in A, which is itself of finite cardinality. On the other hand, V could be of countably infinite cardinality, since it may well be that it is not possible to give a definite and exhaustive list of rules for drawing conclusions about propositions from the existence of records. As far as I can see, P_R^A will also be of finite cardinality, since it seems that only finitely many propositions can be inferred from finitely many records. In any case, this point makes no difference for the argument which follows.

But isn't it too daring, some readers may ask, to make an assumption such as A.3? Doesn't this land us in a problematic naïve realism? My answer is that such an assumption is an essential requirement for scientific investigation and everyday life, since arguably all our knowledge of the external world is based on records contained in substances such as our own brains, books, cameras and such like.[9] Thus, not only can we, but we

[9] Cf. the discussion in C. F. von Weizsäcker, *Aufbau der Physik* (2002), pp. 47–78.

must assume that inference of the type just described is possible, since otherwise we cannot form judgements about the world.

9.3 The Asymmetry of the Local Arrow of Time

9.3.1 The Fixed Past and the Open Future

I will now argue that, on the theory of time laid out in Sect. 9.1, as well as the epistemological assumption of Sect. 9.2, we can account for familiar features of time such as its linear, unidirectional character, as well as the asymmetry between the fixed past and the open future.

To do so, let's consider an ideal recorder A causally interacting with its surroundings, and let's assume, furthermore, that A establishes a local parameter t, as discussed in Sect. 9.1.3. Suppose that A is in some stage $A_{t\alpha}$, where $t_\alpha > 0$. $A_{t\alpha}$ will in general bear records of causal interactions (cf. identification I.1), and being a substance, it can receive more such records (assumptions A.1 and A.2).

Because A counts, records in it are associated with locally measured times t (definition D.6). Furthermore, A counts on the basis of a repeated process affecting it, and so records in the state $A_{t\alpha}$ which we are considering cannot be associated with numbers t greater than t_α. They must therefore be associated with numbers t less than t_α or possibly equal to it (more on this problem below). For each time t less than t_α, there is a set $R^A{}_t$ of records in A. This set will contain at least one record, namely the one associated with the local time keeping process. It may or may not contain further records $r^A{}_t$. On assumption (A.3), from the existence of the records contained in the set $R^A{}_t$, together with the rules of inference v contained in the set V, the truth of some propositions associated with these records follows. We can therefore assign the index t also to these propositions and call them $p^A{}_t$, and the set containing all of them and only them will be called $P^A{}_t$. If, for some t less than t_α, $R^A{}_t$ is empty apart from the record associated with the local time keeping process itself, then, we may stipulate, $P^A{}_t$ will contain information only about the occurrence of time t, and nothing else. In any case, all times t less than t_α will be associated with sets P^A_t, since such times have records, whereas all times t greater than t_α will not be associated with such sets, since these times have no records. This association of local times with sets of records and sets of propositions has three noteworthy consequences:

First, local time measured by A has a linear, as opposed to circular structure. For given two different locally measured times t and t', even if the records r^A_t in R^A_t are indiscernible from the records $r^A_{t'}$ in $R^A_{t'}$, so that two type-identical sets of propositions P^A_t and $P^A_{t'}$ follow from them, P^A_t and $P^A_{t'}$ are nevertheless not identical, because all propositions contained in $P^A_{t'}$ involve $A_{t'}$ rather than A_t. In this way, every local time t is assigned a unique set P^A_t. Thus, even if exactly type-identical interactions take place twice, this does not amount to the recurrence of the propositional content of a past time.

Second, for a state $A_{t\alpha}$ of A, the propositional content associated with a local time t less than t_α, which is given by the set P^A_t, is fixed and unchangeable. Suppose that we are unhappy with a proposition $(p^A_t)*$ contained in P^A_t, and therefore wish to bring about that not-$(p^A_t)*$ is true. This is impossible, because $(p^A_t)*$ is implied by true statements: first, the fact of the existence of certain records which are contained in $A_{t\alpha}$ and which are associated with local time t, and second, the rules contained in set V. Hence, $(p^A_t)*$ is itself a true statement, and so not-$(p^A_t)*$ cannot be true. Nor can we change the propositional content of a local time t less than t_α by attempting to add a proposition to its set P^A_t. For this set, by assumption, contains all and only the propositions which follow from the rules v in V and the existence of the records contained in R^A_t and therefore excludes any other propositions. The notion of adding some proposition, call it $(p^A_t)+$, to P^A_t therefore likewise entails that $(p^A_t)+$ is both true and false. In sum, the local past cannot be changed, because to do so implies a contradiction.

Third, the propositional content of local times t greater than t_α can be influenced. This is because $A_{t\alpha}$ is a state of A, and A, being a substance, can be influenced: it can acquire records through causal interactions, records which in turn generate sets of propositions according to assumption (A.3). However, for $A_{t\alpha}$, the propositional content of a time less than t_α cannot be changed in virtue of the argument in the preceding paragraph.

On the other hand, no such argument applies for times greater than t_α, because no records are contained in $A_{t\alpha}$ for such times. Thus, the local future is open: each time t greater than t_α is not associated with a unique P-set, but can be occupied by infinitely many such sets. The propositional content of such times can therefore be thought of as "branching". In other words, given the basic metaphysical assumption that substances can receive non-essential properties, it is not astonishing that there is a branching set of possibilities which can occur to them. What the philosophy of time needs to explain is not the branching future, but rather why there are no such branching possibilities for the local past, as was done above.

What must be left open at this point is whether t_α itself is occupied by a set of records and a corresponding set of propositions, that is, whether it is the last point of the local determinate past or the first point of the open future. On the one hand, according to definition (D.6), $A_{t\alpha}$ can receive more records. This opens the possibility of several different states of $A_{t\alpha}$, which could be ordered by the relation "before" according to definition (D.3). On the other hand, this is unsatisfactory, because according to identification (I.1), there ought to be a unique state of A associated with t_α. Dilemmas such as this typically plague thought about the present in the philosophy of time. For if the present is associated with some event content, then it seems that it must have some extension. But if it has extension, then there are earlier and later parts of the present, which contradicts the notion of the present as containing only simultaneous events.[10] In other words, the problem arises from associating an infinitesimal point with event content. Operationally, the simplest solution seems to me to view all events associated with the production of records in $A_{t\alpha}$, as well as these records themselves, as belonging to t_α. As of local time t_α itself, there is then no definite fact as to the precise content of the set of records $R^A{}_{t\alpha}$. This set is, so to speak, still under construction, and therefore, so is the propositional content $P^A{}_{t\alpha}$ of t_α.

But what about substances which are not ideal recorders? Do they have a fixed past and an open future? For substances which approximate ideal recorders well, such as ourselves, the structure of time is the same as that

[10] Cf. B. Dainton, *Time and Space*, 2nd edn (2010), pp. 95–102.

for the ideal recorder, but this is complicated by the fact that some records for past local times are deleted and that even preserved records cannot always be accessed. However, as pointed out in Sect. 9.1.1, it is still possible to reconstruct what an ideal recorder would record through comparison with other recorders. In this way, the content of *P*-sets for locally past times can be inferred. Defining time via the ideal recorder therefore has the advantage of explaining the features of the passage of time as we experience it: Past local times *t* are associated with definite and unchangeable propositional content, even though we may not always know that content. In this case, we need to compare with other recorders in order to access it, for example by asking other people or consulting historical records. For future times, on the other hand, there is no definite and unchangeable propositional content. Hence, I submit that such an account serves as the basis for explaining the passage of time which we, even as non-ideal recorders, experience.

9.3.2 How Increase in Entropy and the Passage of Time are Linked

What has been laid out in the previous section can shed light on the difficult problem of the unidirectionality and anisotropy of time's arrow, which is subject to so much debate in the philosophy of physics. Why is it that we can move back and forth in space, but not in time, that we can only act "forward" in time and never "backward"? If we understand time operationally in terms of the ideal recorder, it is the principle of non-contradiction which accounts for this asymmetry: Since past local times are occupied by *P*-sets, the notion of revisiting a past point at time *t* entails that some statement $(p^A{}_t)*$ both is and is not true. Thus, if time moves anywhere, it can only move forward. I therefore propose that the reason for time's arrow is not to be found in irreversible physical phenomena such as the increasing entropy of isolated systems, as many physicists claim (cf. Sect. 6.2). While these phenomena accompany time's arrow, they are not its underlying cause. To use Huw Price's words, they belong not to the "genealogy problem" of temporal asymmetry, but to the "taxonomy problem", that is, the problem of "asymmetries of things *in* time".[11] According to the account which I propose, too, entropy and time are linked, because the former will increase—with overwhelming probability—as the latter does. But the reason for this is that the number of causal interactions increases with local time, not that the irreversible character of entropy increase would "fix" the past. Nor do we need the physically highly implausible concepts employed by cosmic A-theorists, such as a global "now" or ontologically privileged hypersurfaces of simultaneity, in order to account for the passage of time. The tensed structure of time, in short, is a real, but only local phenomenon.

9.3.3 The "Now"

A theory of local time also enables a new approach to the issue of the "now", that is, the question of what the experienced "nowness" of events consists in and whether it is of any ontological significance. It is frequently argued that, since the experienced "now" cannot be illusory, there must be such a thing as "nowness" for the world as a whole (see Sect. 5.1). I propose, however, that we do not need to choose between an absolute, global "now" on the one hand, and entirely dismissing the concept of "now" as illusory on the other. The "now" associated with a state A_t of a recorder A can, for example, be identified with the set of propositions $P^A{}_t$ obtained from the set of records $R^A{}_t$, given a set of rules V. The "now" experienced by us as conscious agents will then be related, but not identical to $P^A{}_t$: as we in some state A_t are affected by the world, and ourselves affect the world, we conclude the truth of some propositions concerning these interactions on the basis of records contained in the set $R^A{}_t$. If we do so correctly, these propositions will be a proper subset of $P^A{}_t$. But $P^A{}_t$ contains many more propositions since, being imperfect reasoners, we cannot draw all conclusions which can be obtained from $R^A{}_t$ and V.

Which records $r^A{}_t$ are produced in A_t is an invariant fact, since the production of these records coincides at the point in spacetime where A_t is located. Hence, given a set V of rules of inference, $P^A{}_t$ is likewise invariant, and unlike the notion of the global "now", does not fall prey to the

[11] *Time's Arrow and Archimedes' Point: New Directions for the Physics of Time* (1996), p. 17.

relativization of temporal concepts brought about by spacetime physics. Thus, the "now" of A_t, unlike the pre-relativistic notion of "all that is occurring in the entire world now", is by no means illusory but has a *fundamentum in re*, namely the content of $P^A{}_t$. Like the distinction between past and future, the "now" too is a real but local phenomenon.

9.3.4 Is Local Time Discrete or Continuous?

Whether time is discrete or continuous is a recurrent question in the philosophy of time and has been debated since antiquity, just as the counterpart question concerning the discreteness or continuity of space.[12] What is the answer to this question according to a local, recorder-based account of time? Before giving such an answer, a few points need to be recalled:

First, the local time count is established on the basis of records of a particular type r^* in a recorder A, of which there can only ever be finitely many, although there is in principle no upper limit to their number. Second, local time units thus established can always be subdivided further by some other type of record. For example, a time count based on records of sunsets can be fine-grained by using pendulum swings. Again, there is no limit to such fine-graining. Thus, for any given rational number, it is always possible in principle to construct a measuring mechanism whose values include that number. There is therefore no smallest "jump", or temporal atom, of local time. It remains true, however, that there can only be finitely many records in a recorder, so that the number of physical records in it can of course never correspond to the full continuum of the rationals. Third, no irrational number (such as $\sqrt{2}$ or π) can ever be reached by such a measuring mechanism, because any number system only permits irrational numbers to be written in the form of infinitely many, non-repeating digits, so that infinitely many records of a particular type would be needed to represent them. Despite this, questions such as "what is the state of A as of π units of local time" or, "which facts obtain at this time" have answers, because such points in local time are covered

[12] For an overview, see, for example, Dainton (2010), Chaps. 16 and 17.

by intervals of rational numbers. The answers can then be given in terms of the smallest such interval. In this sense, we can view irrational points, too, as occupied by content. Fourth, because A in general contains more records than only those associated with the local time counting mechanism, it is always possible for there to be some local time t_α, as indicated by this mechanism, such that there are two—or more—distinct states of $A_{t\alpha}$, as was noted already in Sect. 9.3.1. In this case, there will be a before-order between these states, but it will not be possible to compare their duration among each other, so that we have an instance of *aevum*-like change. We can of course fine-grain the local time counting mechanism further in order to solve this problem, but it can never be completely eliminated.

Where does this leave us? When discussing the discreteness versus the continuity of time, it must be borne in mind that time is, after all, a number, and therefore based on the act of counting. We are not dealing with a substantival time "out there" which could either flow continuously or move forward in discrete jumps. Framing the debate in this way, I submit, misses the character of the object of the debate. According to a local, recorder-based view of time, by contrast, we can answer the question of discreteness versus continuity of time as follows: since the number of records in a recorder is always finite, the ordering of these records is always discrete. Hence, physical time measurement must yield discrete values. However, time, according to definition (D.7), is a parameter established *on the basis of* repeated records, rather than simply the number of the physical records themselves. Because there is no limit to fine-graining this parameter at least in thought, nothing prohibits constructing the full set of the rationals, and even the reals, on any given interval of local time. In this sense, local time can be viewed as a continuum.

10

Objections and Comparisons

Having laid out a proposal for the derivation of "before", of change, of local time, and of its asymmetric character, it is now necessary to answer objections to this proposal and to compare it in more detail with other accounts found in the literature. This will, I hope, both corroborate the proposed account and make clearer what it is and what it is not.

10.1 Charges of Circularity

First, some readers may wonder whether there isn't some hidden circularity involved in the derivation of time's local arrow in Chap. 9. To this, I answer that at no point in this derivation were ontologically primitive and irreducibly temporal notions introduced: It is based on the notion of the set of states of a substance, and an asymmetry on the set of states of an ideal recorder, all of which are atemporal notions. These allowed us to define "before". Nor were underived temporal notions required in the account of change, of counting in terms of nested sets based on r^*, of local measurement of duration, of the association of records with locally

D. Saudek, *Change, the Arrow of Time, and Divine Eternity in Light of Relativity Theory*, Palgrave Frontiers in Philosophy of Religion,
https://doi.org/10.1007/978-3-030-38411-1_10

measured times, or of the fixity of the local past—as opposed to the openness of the local future—through the principle of non-contradiction.

It is illustrative, at this point, to consider an objection to the present account based on the notion of a "deleter": a substance *D* which starts off with a set of *n* records and successively deletes them, one by one, as time moves forward.[1] If we apply definition (D.3) in Chap. 9 to it, we find the same asymmetries in the existence of states of *D* as we did in the states of the ideal recorder, but in the reverse temporal order. It seems, therefore, that temporal succession is insufficiently determined by definition (D.3).

To answer this objection, consider deleter *D* as it transitions from a state where it has *k* records to one where it has *k-1* records, where *k* is a natural number such that $n \geq k \geq 1$. If we picture this scenario before our mental eye, we start with an impression of $Dx_1 \ldots x_k$, to use the notation introduced in Sect. 9.1.3 and then have an impression of $Dx_1 \ldots x_{k-1}$. But this means that there is a state of ourselves with a record corresponding to $Dx_1 \ldots x_k$, and another which contains both a record corresponding to $Dx_1 \ldots x_k$ and to $Dx_1 \ldots x_{k-1}$, but no state of ourselves containing a record of $Dx_1 \ldots x_{k-1}$ without $Dx_1 \ldots x_k$. This asymmetry in the set of our states, by definition (D.3), yields the correct before-order whereby $Dx_1 \ldots x_k$ is before $Dx_1 \ldots x_{k-1}$. In other words, it is not possible to consider the deleter successively losing records without accumulating—at least in thought—records of it doing so. If we ourselves lost records in parallel with *D*, it would not be possible to judge that *D* lost records. This judgement therefore requires a recorder sufficiently close to an ideal one, and the criterion in (D.3) applied to this latter recorder suffices to define the before-order also for *D*. The deleter is thus simply a special case of a substance losing properties, as defined in (D.4) in the previous chapter. Definition (D.3) is thus sufficient to define temporal succession.

Finally, some may object that if substances causally affect each other, this cannot be conceived of without time, so that an underlying time must be presupposed. To this I answer that, as argued in Sect. 8.2, neither does causation need to have an intrinsic duration in order for causal processes to be measured to take nonzero time nor are the states before and after a causal interaction essential to the interaction itself. In addition, as

[1] I thank Markus Seidl-Nigsch for providing this thought experiment.

was pointed out, the electromagnetic interaction plays a central role in causal interactions between substances, and this interaction does not seem to be intrinsically temporal.[2] We may therefore outline the relationship between electromagnetic causal interactions and time as follows: the electromagnetic interaction is responsible for the existence of different states of substances, which in turn gives rise to time. Contrast this with the view whereby time would be required for the electromagnetic interaction, which then gives rise to different states of substances.

Nevertheless, there is a certain truth in the adage "no change without time": if there is a substance *B* which has at least two states, and a substance *A* sufficiently similar to an ideal recorder—in many cases, even a very simple recorder will do—is affected by *B*, then quantification over *A*'s states will yield a before-order, so that we have an instance of change. In addition, *A* can always in principle quantify this change, provided that it is equipped with some counting mechanism. In this sense, wherever there is change, there is time, so that time is indeed necessarily a concomitant of change. But this does not mean that it should be conceived of as independent of change, as Kant thought.

In sum, the general objection whereby deriving time in terms of change must involve circularities, together with the specific objections under this heading given in Sect. 7.3, can be considered answered.

10.2 The Proposed Causal Theory does not Rely on Causal Asymmetry[3]

A further clarification should be added at this point: the theory of time laid out in Chaps. 7, 8 and 9 can be called a causal one because, according to it, causal interactions between substances account for time. This needs

---

[2] Again, the term "substance" is taken to mean a relatively macroscopic, composite object, as opposed in particular to a subatomic particle (cf. the discussion in Sect. 3.5). Interactions which result in intrinsic change of substances are by and large electromagnetic. Tidal forces—which however are likewise associated with a zero spacetime interval— are perhaps the most relevant exception.

[3] This section incorporates material from my article "Beyond A-theory and the block universe: A non-circular derivation of 'before', change, and the local arrow of time", as cited at the beginning of Chap. 9.

to be distinguished from the claim, often associated with the term "causal theory of time", that some form of asymmetry between causes and effects accounts for temporal asymmetry. Several different models which employ this basic strategy can be found in the literature: for example Hans Reichenbach's mark method,[4] D. H. Mellor's reduction of temporal to causal order[5] or Mathias Frisch's argument whereby temporal asymmetry can be understood in terms of causal and explanatory models.[6] A related account is that developed by Adolf Grünbaum, where the relationship of temporal betweenness of events is explained in terms of an event's being necessary for the causal connectedness ("*k*-connectedness") of two other events.[7]

By contrast, I do not claim that temporal order follows from causal order or that it is derived from an asymmetric relationship between cause and effect. Effects are not later than their causes because they depend on their causes. Rather, I claim that different states of substances exist due to causal interactions and that time can be constructed on this basis by quantifying, timelessly, over states of a particular type of substance, namely an (approximately) ideal recorder. This leaves open whether such states also stand in an asymmetric relation of causal dependence among each other, and in which way events or states of affairs stand in such relations. These questions are clearly of fundamental philosophical importance, but their resolution, I claim, is not necessary in order to establish a non-circular theory of local time.

10.3 The Charge of Subjectivism

Some may object that the account of time's arrow in Chap. 9 is too "subjective", since it is based on a recorder as a subject, albeit not necessarily a conscious one. To answer this objection, two meanings of the term

---

[4] H. Reichenbach, *The Philosophy of Space and Time* (1958), pp. 135–138, as cited in A. Grünbaum, *Philosophical Problems of Space and Time,* 2nd edn (1973), p. 180, which see for a discussion.

[5] *Real Time II* (1998), pp. 105–117.

[6] "Time and causation", in A. Bardon, H. Dyke (eds.) *A Companion to the Philosophy of Time* (2013).

[7] Grünbaum (1973), pp. 193–197.

"subjective" must be distinguished: first, in the sense of "involving a subject"; second, in the sense of "susceptible to subjective interpretation". The first type of "subjectivity" is arguably involved in any measurement of observables of an object, such as length, mass, or spin. This is because measurement involves the comparison of an object with a standard,[8] and such comparison involves an interaction between a subject—again, not necessarily a conscious one—and the system. This, however, does not lead to "subjective" measurement outcomes in the second sense, because measurement, given an operational procedure—which in a relativistic context also includes a particular frame of reference—yields a unique result which leaves no room for subjective arbitrariness.

It is often overlooked that both classical and quantum mechanical measurement share the above features. That is, both are "subjective" in first sense, but not in the second. Neither is classical measurement subject-less, nor does quantum mechanics allow a subjective, arbitrary choice of results in a way which appeals to those sceptical of scientific objectivity. In both cases, a subject interacting with nature obtains unique answers which nature, so to speak, "throws" at the subject, that is, "objects" to it in the etymological sense. Of course, important differences between classical and quantum mechanical measurement remain, in particular the irreducibly statistical character of the latter as opposed to the former.

What is true of classical and quantum mechanical measurement also applies to local temporal measurement. It is "subjective" only in the first sense, because it involves a recorder as a subject, but not in the second: given an ideal recorder, there is a unique before-order between any two records, and to the local events to which they correspond. Hence also, given a local counting mechanism, which is itself based on records, the time index of any record (according to definition (D.6) in Chap. 9) is unique. In addition to this, both the local before-order and the local time index are relativistic invariants: given a recorder *A* equipped with a time-keeping mechanism, all observers independently of their frame of

---

[8] A point in consonance with the general approach of the Erlangen school, as exemplified in P. Janich's *Das Maß der Dinge: Protophysik von Raum, Zeit und Materie,* 2nd edn (2016).

reference will agree which order between records and corresponding events *A* measures, and hence also which time index *A* assigns to them.

10.4 The Objection from General Relativity

It can furthermore be objected that the very notion of causal interaction, on which my account of local time rests (by assumption A.2 in the previous chapter), is incompatible with general relativity (GR). This objection can roughly be summarized as follows: The notion of causal interaction relies on the transfer of energy. But statements about the transfer of energy from one system to another are only warranted if an integral conservation law applies, thereby allowing us to keep track of where in space, and by which systems, energy is gained or lost. As Erik Curiel succinctly puts it:

> The fact that it is always the same *amount* of energy that is gained and lost by interacting systems is supposed to preclude the idea that energy can be created or destroyed, and so warrant the inference that energy is actually *transferred* between interacting systems, as required.[9]

Curiel then shows that such transfer accounts cannot be formulated in general relativity (GR), because only a differential (and hence, in a curved spacetime, covariant) conservation law, but no integral conservation law applies in a general relativistic spacetime.[10]

I agree with Curiel that GR imposes constraints on physicalist accounts of causation. However, it cannot be used to undermine all notions of causal interaction. The main reason for this is that GR is fundamentally a metric theory. Very roughly speaking, its fundamental equation, the Einstein Field Equation, tells us how certain measurable quantities, spatial and temporal ones, relate to the quantities contained in the stress-energy tensor. But measurement is itself a type of causal interaction, either among the parts of an object (as in the case of a ticking clock) or of

[9] "The constraints general relativity places on physicalist accounts of causality" (2000), p. 20.
[10] Ibid., pp. 20–28.

an object with its environment (as in the case of length measurement). Therefore GR, far from ruling out causal interactions, presupposes them, not merely as an essential part of the context of discovery of the theory, but in the sense that terms fundamental to the theory cannot be understood apart from such interactions. This is not to deny the fundamental character of GR as a physical theory, which Curiel rightly draws attention to. As far as we know, everything obeys the quantitative predictions of GR. However, the theory is not metaphysically fundamental, insofar as it is based upon concepts defined independently of the theory itself.

What might be an account of the notion of causal interaction which *is* compatible with GR? It is beyond the scope of this book to give a full-fledged account of causation. It suffices to show that accounts are possible which do not depend on integral conservation laws. An example is that given by Mumford and Anjum, for which I already stated my preference in Sect. 8.2. In it, causal interactions are thought of as point-like, so that we do not need to keep track of portions of energy over space and time in order to support a causal claim. To use again the example of two colliding billiard balls—let's call them *a* and *b*—it is "the impact which is causally efficacious, not how ball *a* got to the point of impact".[11] This being so, a necessary condition for the causal claim that *b* acquired energy from *a* is that a differential equation of energy conservation—which in a general relativistic spacetime will have a covariant form—is satisfied at the point of impact, whereas an integral one which ensures that energy is conserved over a spatiotemporal volume is not necessary. It may be objected that the collision event between *a* and *b* cannot be viewed as strictly point-like, but is itself extended in space and time. However, this event can be analysed in terms of smaller events involving the emission and absorption of photons by the molecules contained in *a* and *b*, for which it remains true that these events, not what happens before or after them, are causally efficacious. Hence, an account of causation such as Mumford and Anjum's is not undermined by the absence of an integral conservation law in GR.

Suppose, in a next step, that even a differential conservation equation failed to hold. For example, let's imagine that, in a collision between two

[11] See Sect. 8.2 for the full quote and reference.

point-particles *a* and *b*, where *b* is initially at rest, the total momentum miraculously changes, so that *b* flies off with more momentum than *a* had before impact. Clearly, this would preclude the causal claim whereby *b*'s momentum came entirely from *a*. But even then we would be justified in applying assumption (A.2): *b* acquired its momentum by interacting with its environment, at least in the minimal sense of something distinct from *b* itself, which in this case is the assumed miraculous source of the extra momentum.

In sum, assumption (A.2) is in no way undermined by GR. The theory does not rule out the very notion of causal interaction, but only particular types of physicalist accounts thereof.[12]

10.5 Comparison with Janich's Protophysics of Time

The account of local time which I propose is consonant with the operationalist approach of the Erlangen School, and in particular Peter Janich's protophysics of time,[13] in the sense that time is understood in terms of a quantification of change with the help of a repeated standard change. The principal difference lies in the derivation: Janich starts his account of time measurement with change and motion as notions known by example from everyday experience[14] and therefore not in need of further analysis. After a detailed geometric discussion of comparisons between different locomotions,[15] Janich then introduces repeated change, also known by example, as the basis for quantification of change.[16] From this, Janich provides rigorous, and in my view helpful, definitions of time measuring

[12] Similarly, Curiel points out that his critique does not apply to all accounts of causation, but only ones whereby "causality is a *physical* relation holding among *physical* entities" and which "take the transfer (and the propagation) of energy to play the most important role in constituting the causal relation" (p. 2). I went further in arguing that even a transfer account is viable, as long as it is based on point-like interactions. But this does not exclude that other accounts of causation, not based on the transfer of energy, may be viable.

[13] *Die Protophysik der Zeit*, 1st edn (1980).

[14] p. 142.

[15] pp. 151–179.

[16] p. 192.

devices and of their regularity and constancy, based on a distinction between a device itself and a part of it which functions as a time indicator.[17] The relations "earlier" and "later", as well as the asymmetry between past and future are, according to Janich's account, consequences of time measurement so introduced, and therefore do not stand in need of a further derivation.[18]

My proposal, by contrast, is to derive the definition of "before" from changeless notions, and from there on the notions of change and of a repeated standard change as the basis for time measurement, as carried out in Sects. 9.1.1–9.1.3. Also, I have proposed a derivation of the irreversibility of time in Sect. 9.3.1 based, not on measurement alone, but on propositions and their truth values. Janich's account of clocks and chronometry is much more detailed than my rather minimalistic proposal of defining time in terms of several instances of a property $r*$ in an ideal recorder A. But the two approaches for time measurement can be combined, for example by defining repeated instances of records of the same type $r*$ in A as records of a time measuring device, as construed by Janich, in A's vicinity.

10.6 Comparison with the IGUS Model of Temporal Asymmetry

The physicist James B. Hartle proposes that the asymmetry of time is best understood in terms of IGUSes, that is, "information gathering and utilizing systems".[19] This approach is consonant with what I have laid out in Chap. 9, where the difference between past and future was explained in terms of records contained in a substance and propositions derived from them. There are, however, important differences: Hartle's account presupposes, first, a time interval in which the IGUS erases old images in its memory and acquires new ones,[20] and second, a direction of time defined

[17] pp. 194–210.
[18] pp. 214–215.
[19] "The physics of 'now'" (2008), p. 1.
[20] p. 2.

by the increasing entropy of the universe.[21] Hence, a pre-existing spacetime structure in which the IGUS is set is assumed. In Craig Callender's words:

> Is this [the IGUS] enough for passage? I think it probably isn't. Ultimately with IGUS we have a robot instantiating various asymmetric processes and maintaining an asymmetric correlation structure with the outside world ... [However] we don't have movement yet, the *whoosh* and the *whiz.* Nothing seems to "crawl up" the IGUS's worldline, thereby making time flow ... The memory asymmetry doesn't provide us with our desired feature of *something moving through time.*[22]

The account which I offered, if successful, solves this problem by deriving the notions of "before", of change, and of the arrow of time in terms of the recorder itself. Thus, it does not need to presuppose an existing temporal structure in order to explain the temporal asymmetry "experienced", so to speak, by the IGUS. A further difference is that, according to Hartle's account, it is in principle possible for a robot to organize its memory in different ways relative to the direction of time, for example, it could also remember its future.[23] By contrast, I have proposed reconstructing the familiar notion of the "past" through records contained in a recorder *A* and the propositions which can be derived from them, and the "future" in terms of *A*'s ability to acquire more records. Hence, according to my account, a recorder cannot in principle remember its future.

---

[21] pp. 6–8.
[22] C. Callender, *What Makes Time Special?* (2017), p. 247, emphases in the original.
[23] *Op. cit.*, pp. 8–9.

11

Our Changing World

11.1 Beyond A-theory and the Block Universe

After all these detailed investigations in physics and philosophy, it's about time to try to see the wood, rather than the trees, and to outline, in general terms, a proposal for a relativistically correct philosophy of time and change. This will pave the way towards an investigation into the relationship between divine eternity and the changing world, in Chaps. 12 and 13. Based on the groundwork undertaken so far, many aspects of the phenomenology of time, as known from everyday experience, can be saved and should not be declared to be illusory. However, some deep-seated intuitions about time and its passage must be overcome.

First, change is real, that is, things in spacetime acquire or lose properties by interacting with each other. Spacetime physics does nothing to undermine this notion. Since the existence of change does not, as Gödel thought, depend on a mind-independent lapse of time with its moving "now", but rather, as I argued, time arises from change, the fact that relativity of simultaneity undermines the moving "now" does not undermine the reality of change.

D. Saudek, *Change, the Arrow of Time, and Divine Eternity in Light of Relativity Theory*, Palgrave Frontiers in Philosophy of Religion,
https://doi.org/10.1007/978-3-030-38411-1_11

Also, we can affirm the reality of true becoming. Within spacetime there are instances of causation which make possible events actual, so that the content of all of spacetime—if we choose to model it as a four-dimensional block—is not what it would be if these instances did not occur. A necessary condition for their existence is that they not be predetermined by the past light cone at the particular points in spacetime where they occur. We might say, in other words, that the story of the universe is constructed to some extent from within rather than only from laws and initial conditions. Agent causation is an example of special importance to us as conscious beings who, as we have every reason to believe, can make a genuine difference to the content of spacetime and the development of the universe, however slight this difference may be in comparison with the causal impact of incomparably more vast entities existing in it.

I submit that the local past along a world line is fixed, and the local future open since, if time is understood operationally in terms of recorders, there is a unique and unchangeable *P*-set for any past time, but infinitely many possibilities for the content of the *P*-set of any future time. Therefore, the most fitting metaphor for this model of local time is not the tree, whose branches separate at the present and represent the possible future courses of events, but rather something like a horsetail, a plant where there is branching from every node above a certain threshold. Moreover, the "now", as known from everyday experience, is not illusory but has a *fundamentum in re*, as argued in Sect. 9.3.3. This picture of local time asymmetry matches our everyday experience as conscious agents: at any stage of our existence, we can actualize states of affairs and can only meaningfully deliberate about actualizing future ones, or possibly present ones, whereas the past cannot be revisited in principle.

Evolutionary cosmology and the notion that the world as a whole undergoes a development from earlier to later stages can likewise be upheld, at least insofar as these stages can be put into a total and directed order, or B-series, by using cosmic time, which is uniquely suited for a unified causal account of the development of the universe, as described in Chap. 4. This is the case for the known universe, but it is not known whether cosmic time also applies beyond our particle horizon.

However, we have every reason to discard the notion of a global "now" and of the set of all events happening in the universe now. Questions about what is happening now in the Andromeda galaxy, or even in the world as a whole, have no meaningful answer, because they are based on the extrapolation of the concept of "now" to contexts where it simply does not apply. In this sense, such questions are similar to asking where the "downward" direction is in an environment where the gravitational field is zero. The history of the cosmos is therefore not an A-series.

Any notion of a substantival passage of time can likewise be discarded. What it is meant to explain—change, the duration of processes, the one-way street character of time, or the difference between the open future and the fixed past—can, in my view, be accounted for without it. Assuming a substantival passage leads to unnecessary problems concerning what its causal role might be and to equally unnecessary conflicts with physics about the location of the moving "now" in spacetime. Furthermore, with substantival passage we inevitably run into contradictions about the temporal properties of events, as McTaggart showed.

All events which actually occur in spacetime have, according to the account I propose, the same ontological status. As agents in spacetime, we must distinguish between past and future events: past ones can affect us-now and we-now can affect future ones. But this does not mean that past, present, or future events differ in their mind-independent ontology. As far as the latter is concerned, we need only to distinguish between events which occur from those which do not, between possible events and those which are actualized. It bears repeating that all this does not entail a block universe view whereby all events in spacetime are "already there", since this impression is based on an erroneous extrapolation of the familiar notion of "already" to spacetime as a whole. Nor, for reasons already given, does the equal ontological status of past, present, and future events preclude that agents have an open, branching future.[1] In any state of their existence, agents have several options to act, but for an agent's past state, there is a unique *P*-set describing the causal interactions between the agent and her environment, which by the principle of

[1] On this point, I disagree with D. H. Mellor, according to whom the equal ontological status of past, present, and future events make a branching future impossible (*Real Time II* [1998], p. 125).

non-contradiction cannot be changed. Since events in our past are fully real, they themselves serve as truthmakers for propositions about them, so that we need no substitutes, such as presently existing traces of past events or patches of spacetime, to act as truthmakers.

Time itself is neither an object existing "out there" in reality, nor a mere mental construct, but, to use a scholastic phrase, an *ens rationis cum fundamento in re*, that is, an entity of reason with a foundation in reality. It is not ontologically irreducible, since it can be derived from timeless notions, but what is not irreducible is not on that account illusory. However, time is in a certain sense all-pervading, because change is, and—in the sense given in Sect. 10.1—wherever there is change, there is time.

11.2 Fixing the *K*-relation

At this point, we can revisit the issue of causal asymmetry in spacetime, left open in Sect. 3.1.2. Why are causally connectable events related by an asymmetric *K*-relation in such a way that their time order cannot be switched, given that their causal connectability is time-symmetric, that is, is preserved if the temporal separation t in the spacetime interval is replaced by $-t$? As far as the special theory of relativity (SR) is concerned, the answer is simple: if x and y are two causally connectable events, and y is measured to occur before x in one frame of reference, the speed of light as the absolute speed limit in the universe precludes the existence of frames of reference for which x occurs before y. In a theory of causation by which causes precede their effects, this fact preserves causality in the universe, because it ensures that they invariantly do so in all frames of reference. However, as argued in Sect. 8.2, causes are strictly speaking simultaneous with their effects, although causal processes can still take finite time. But also in this alternative account of the relationship between time and causation, it turns out that the speed of light as an absolute speed limit ensures a consistent universe. I will argue that, if two events x and y are causally connectable and if, for some observer A, y is before x, then all observers must agree that y precedes x, not only because no frame of reference can move faster than light relative to another, but because to assume that there is an observer C for whom x is before *y* leads to contradictions. Let's take "$B_A\, xy$" to mean "*y* is before *x* for observer *A*". The argument then runs as follows:

1. *x* and *y* are causally connectable events, that is, the spacetime interval between them is greater than or equal to zero.
2. If two events are causally connectable, at least one of them can influence the probability of the other occurring.
3. If the occurrence of an event is certain, then the probability of its occurring cannot be influenced.
4. If an observer is reliable, he or she forms true judgements at any stage of his or her existence.
5. *A* and *C* are reliable observers.
6. *A* and *C* disagree about the temporal order of *x* and *y*.

Let's say that, from *A*'s perspective, *y* is before *x*. Thus, there is a time *t*, as recorded by *A*, such that *y* is certain, but *x* is not ($B_A\, xy$). Then, the probability of *y* cannot be influenced by *x* (from 3). Therefore, it must be true that the probability of *x* can be influenced by *y* (from 2). Because *A* is reliable, these judgements are true.

From *C*'s perspective, *x* is before *y*, so that we get the reverse of the above: there is a time *t*, as recorded by *C*, such that *x* is certain, but *y* is not ($B_C yx$). The probability of *x* cannot be influenced by *y*. Therefore, the probability of *y* can be influenced by *x*. Again, *C* is a reliable observer, so her judgements are true.

From these assumptions, we get two contradictions: the probability of *y* both can and cannot be influenced by *x*, and the probability of *x* both can and cannot be influenced by *y*.

Hence, if two events are causally connectable, it is logically (not just physically) ruled out that two reliable observers can disagree about their sequence in time. $B_A\, xy$ therefore implies $B_O\, xy$ for any observer *O*. We can then drop the subscript and write simply *Bxy*.[2]

---

[2] The basic move of arguing from probabilistic considerations to temporal asymmetry which I have adopted in this section is inspired by Mellor (1998), ch. 12, in which he argues against causation backwards in time. Mellor's approach is quite different from the one I propose, since he assumes that causes precede their effects and that events have no branching futures. Nevertheless, the results are quite similar.

This allows us to fix the *K*-relation by defining that *Kxy* is true exactly if *x* and *y* are causally connectable and *Bxy* is true. This relation is asymmetric: if *x* and *y* are causally related and *Bxy* is true, then *Byx* is not true, and hence nor is *Kyx*. It is transitive, since if *x* and *y* are causally connectable, and *y* and *z* are causally connectable, and if *Bxy* and *Byz*, then *x* and *z* are causally connectable and *Bxz* is true, so that also *Kxz* is true. It is furthermore irreflexive, because *Bxx* is true for no event *x*.

This amended *K*-relation therefore reduces to the familiar one introduced in Sect. 3.1.2. Note, however, that in order to define it, it was not necessary to assume that causes precede their effects. Rather, the amended *K*-relation relies on the temporal asymmetry as experienced by observers, involving the certainty of recorded events and the openness of any event's future, and on the fact that the causal process of which *x* and *y* are parts is extended in time.

11.3 Local Time and the Gödel Universe

The possibility of a Gödel universe which includes closed timelike curves (see Sect. 4.2) seems to contradict the result of the previous section, because in such a universe, an event *y* can be in both the past and the future of another event *x*. In addition, it could be claimed that said possibility also contradicts the result, obtained in Sect. 9.3.1, whereby local time cannot have a circular structure. How, given this, does local time relate to the Gödel universe which, after all, is permitted by general relativity, and therefore cannot simply be dismissed out of hand as absurd?

The answer is simple: while time as measured locally does not permit revisiting the past, this does not prohibit time as measured by a coordinate observer to have a circular structure. This structure is then a metric property of a spacetime, but it cannot reflect the passage of time experienced by a local observer with its characteristic asymmetry between past and future. Thus, it cannot be concluded from general relativity that it is possible to travel to the past and, for example, kill our own ancestors. To do so would be to confuse a metric property of a spacetime with time as measured locally. This confirms Steven Savitt's observation whereby "time *qua* passage is a *local* phenomenon, tied to a world line". If, in addition, passage is essential to time, then we can take the possibility of the Gödel universe not as an argument against the passage of time as such, but rather as confirmation of James Harrington's dictum whereby "proper time *just is time*".[3]

[3] See Sect. 5.4 for the sources of both quotes.

Part II

God, Change, and Time

12

Does God Change?

What has been developed so far can now be applied to questions surrounding God's relationship to the changing and temporal world. The contemporary debate about this relationship has produced an extensive body of literature: First, a number of analytic philosophers have investigated issues such as whether God's eternity is compatible with His action and personal involvement in an evolving world or whether divine foreknowledge of future contingents can be reconciled with human freedom. Second, scholars working at the interface between science and theology have explored how scientific insights on time relate to theology and vice versa. In the latter field, the debate has concentrated on the crucial question of whether we are living in a truly dynamic, changing world or rather in a timeless block universe. In my view, this debate has been marked by a too strong polarization between A-theory and the block universe, as though these were the only available alternatives. I have argued in favour of a third option: a model which recognizes the possibility of true becoming without thereby falling into the paradoxes of classical A-theory. This model was made possible through an Aristotelian understanding of time. In the following two chapters, I will argue that it can be used to make

D. Saudek, *Change, the Arrow of Time, and Divine Eternity in Light of Relativity Theory*, Palgrave Frontiers in Philosophy of Religion,
https://doi.org/10.1007/978-3-030-38411-1_12

headway in answering the philosophical questions about God and time just mentioned.

12.1 The Ground of Being as Limitless Being

However, before doing so, we need to take a step back. What reason is there to bring in God in the first place? The main *philosophical* motivation for assuming the existence of God is to provide an ultimate explanation for the existence of contingent things, including the universe as a whole. There are of course other motivations for this assumption, such as personal experience and intuition or authority and tradition. However, as epistemic bases to the claim that God exists, these motivations suffer from serious drawbacks: Personal experience and intuition are not, or only with great difficulty, accessible intersubjectively and are not as such susceptible to rational analysis. Authority and tradition, while indispensable for the acquisition of knowledge, do not by themselves suffice to justify a truth claim, since such a justification would be self-referential. I am therefore critical of reformed epistemology, which appeals to the *sensus divinitatis* and the epistemic rights of the believer in order to establish a truth claim about God as justified.[1] In scientific or philosophical investigation aimed at discovering true propositions about reality, we do not merely demand that the proponent of a claim be in their epistemic rights to advance it, important as this point is. Rather, we ask whether a particular truth claim has a *fundamentum in re* to justify it. This is not to say that all well-meaning participants in the debate will agree with what follows, but rather that philosophical theology should attempt to make its case on the basis of rational argument.

Arguments from contingent and limited beings to an absolute and limitless ground of being have of course been made in a number of ways, and it is beyond the scope of this book to assess and compare these.[2] For the sake of our subsequent investigations about God and time, I will

[1] See, for example, A. Plantinga, *Warranted Christian Belief* (2000), ch. 6.

[2] For a systematic overview, see, for example, W. Löffler, *Einführung in die Religionsphilosophie*, 2nd edn (2013), ch. 3.

present the essentials of an argument from contingency, following an approach developed by William N. Clarke, which is itself inspired by the work of Thomas Aquinas.[3] I will then argue that the ground of being must be thought of as personal.

Contingent entities—whether events, things, systems of things, or states of affairs—do not explain themselves, but rather require something beyond themselves as the foundation and explanation of their existence. This applies to any entity or system of entities considered in isolation, but also to the universe as described by cosmology. Not surprisingly, a great variety of explanations for the physical universe have been proposed: It may have arisen from a random quantum fluctuation in a pre-existent scalar field. Our world could be a computer simulation designed by forms of intelligence more advanced than ourselves.[4] It could be explained immediately by divine design. Or it could be a member of a large, perhaps infinite, family of spatiotemporal bubbles which together constitute a vast multiverse, within which new baby universes are constantly produced, and which is itself governed by some set of laws, for example, those of quantum mechanics. Exotic as some of these scenarios may be, none of them is patently self-contradictory, so that all of them remain live options.[5]

Do these proposed explanations for the physical universe themselves require an ontological foundation beyond themselves? The pre-existent scalar field and the laws governing its behaviour, the aliens running computer simulations, or the multiverse governed by laws describing it, are likewise contingent beings. That is, they realize only a certain set of possibilities at the expense of others, could have been other than they are in fact and might not have come into existence at all. Therefore, these entities likewise require an ontological foundation beyond themselves, as does any form of limited being which might in turn provide a foundation for their existence.

[3] W. Norris Clarke, "Is a natural theology still possible today?", in R. J. Russell, W. R. Stoeger, G. V. Coyne (eds.) *Physics, Philosophy and Theology* (1988).

[4] Nick Bostrom argues that this hypothesis is likely to be true in "Are you living in a computer simulation", *Philosophical Quarterly* (2003).

[5] A fine survey of similar proposals is given by P. C. W. Davies in *The Goldilocks Enigma* (2006). See especially pp. 295–303.

This applies also if we suppose an infinite causal chain in order to explain the existence of a particular entity. Consider, for example, a world of infinite age in which horses have always existed: every horse in this world is begotten by other horses, so that each horse is sufficiently explained. This, however, does not explain why the infinitely old world of horses exists in the first place. In other words, such a world is likewise contingent, realizing some possibilities rather than others, and therefore requires a foundation of its existence beyond itself.

Since contingent entities exist and require a foundation beyond themselves, since no contingent entity can provide the foundation for its own existence (although it could provide one for other entities) and since an infinity of causally connected contingent beings cannot provide a solution, because it is itself a contingent being, the only way out is to assume the existence of a non-contingent entity as a foundation for everything else which exists. Such a being must be truly limitless, containing all possibilities within itself and possessing all positive properties.[6] As Clarke explains:

> *Any being self-sufficient for its own existence must be infinite in perfection*, i.e. unlimited in its qualitative fullness of all perfections. Or: *No self-sufficient being can be finite*. Why? Let us suppose it were finite. This means it would be one determinate, limited mode of being (limited in qualitative intensity of perfection) among at least several other possible modes such that at least one higher mode were possible … But no finite being can select its own essence and confer existence on itself. For then it would have to pre-exist its own determinate actual existence (in some indeterminate state), pick out what it wills to be, and confer this upon itself. All of this is obviously absurd, unintelligible. It follows that no determinate finite being can be the self-sufficient reason for its existence as *this determinate being*.[7]

[6] The notion of "positive property" which I am employing here plays a key role in Gödel's ontological argument for the existence of God. However, the contingency argument presented here is not an ontological one like Gödel's, since it starts from the existence of limited beings as an *explanandum*, rather than with the notion of a perfect being, a notion which, according to the proponents of ontological arguments, implies the actual existence of such a being.

[7] Clarke (1998), pp. 114–115, emphases in the original.

The objection “what causes the ground of being” appears therefore to me to be misplaced, since it overlooks the distinction between limited, contingent being, which cannot provide the foundation of its own existence, and the full perfection of unlimited being. To this latter, the questions “why is this entity such rather than other?” or “why was this possibility realized, rather than some other?” do not apply, since it contains all possibilities and possesses all positive properties.

Since the limitless ground of being must be such as to provide an ontological foundation for everything in the universe, nothing in the universe can realize possibilities not already contained in this ground of being, or exceed it in any positive property. Rather, entities different from the ground of being will necessarily realize only some, but not all possibilities contained within the ground itself. They can therefore be said to participate in its perfection, while lacking the fullness of this perfection.

A consequence of the above is that the ground of being must be thought of as possessing the properties associated with persons, in particular those of knowledge and volition, at least insofar as these properties are positive. In human beings, both knowledge and volition contain negative elements: For example, knowledge of what is the case can often only be acquired by excluding what is not the case. Volition is most often, though not always, associated with the fulfilling of a need, and hence a defect. The fact that negative elements are associated with both knowledge and volition does not however remove the positive content of either, but rather presupposes such a positive content: Knowledge of an entity cannot be reduced to the exclusion of what that entity is not. In meeting or providing for needs, we wish the positive existence of ourselves or of other beings. By the principle stated in the last paragraph, this positive element must also be found in the ground of being, albeit not in a limited state, but rather in infinite perfection. Volition in the ground of being can therefore not be understood as desire, since the ground of being is already in possession of all positive properties and lacks nothing, but rather as willing the existence and good of beings.

Having argued that the unlimited ground of being is knowing and willing, and hence personal, I will henceforth employ the term “God”, rather than “ground of being”, in accordance with everyday language.

12.2 Divine Action

However, the concept of God as an unlimited and perfect being with all positive properties and affected by no contingency seems, at first sight, to place us before a curious dilemma: If God is the ontological foundation and the ultimate cause of all contingent beings, He must somehow interact with them, at least in order to create them and sustain them. In addition, theism has traditionally held that God acts in the history of the universe in special ways which go beyond the creation and sustaining of creatures. Does God change as a consequence of any of these types of interactions? If so, how can He be said not to be affected by contingency? If not, how can He interact with the world? Many participants in the current debate have rejected the notion of an unchanging God as incompatible with that of a God who is personally involved with His creatures. Thus, John Lucas argues that, if God were changeless, He could not act in history.[8] Similarly, William L. Craig holds that God's decision to create a world has made Him subject to change since then.[9]

However, the notion of God's being subject to change is in conflict with His being the ground of all which exists. For if God changes due to His interactions with contingent, finite beings, by definition He acquires or loses some positive property p, where the nature of p depends on the contingent beings involved and on the type of interaction with them.[10] Then, there are two states of God, one with p and one without it. But this is impossible, since God without p is not a being in possession of all positive properties, is therefore limited and does not qualify as the ground of being. Such a ground of being containing all positive properties is however required given that there are limited beings, as argued above. Thus, even a spiritual being of extremely great power, if it were subject to change, would require an ontological foundation beyond itself, and would therefore not be God, at least not in the sense of an ultimate ground of everything which exists. The existence of such a great spiritual

[8] J. Lucas, *The Future: An Essay on God, Temporality, and Truth* (1989), pp. 214–216.

[9] *God, Time, and Eternity* (2001), p. 69; and *Time and the Metaphysics of Relativity* (2001), p. 237.

[10] Cf. the explication of change in Sect. 7.2. The argument that God as a perfect being cannot be subject to change was developed by Plato in *Republic*, 381.

being is of course possible, but a philosophical argument in favour of its existence, as laid out in the previous section, would no longer be available, since such an argument leads to an entity characterized by all positive properties. But then, it seems that we have little other epistemic motivation to assume its existence other than personal experience, authority, or cultural tradition.

But can divine action be reconciled with the notion of an unchanging God? A first step in doing so is to analyse what it means for an entity to act upon another. Clarke reminds us of the scholastic adage whereby *actio est in passo*. If *A* acts upon *B*, *A* causes *B* to acquire or lose some property *p*. That is, for the proposition "*A* acts upon *B*" to hold true, it is required that there is *p*-change in *B* (for some property *p*) and that this is due to the causal influence of *A*. By itself, this definition says nothing about the way in which *A* changes because it acts upon *B*, or indeed whether it changes at all. But of course, in all causal interaction known from everyday experience or scientific investigation, both partners are affected, so that the action is also *in agente*. This is most obvious in interactions involving direct physical contact between bodies: since each body occupies a limited region of space, one body can only affect another by undergoing some change itself, which will at least involve a change of the relative position of its own parts. But the same principle also applies to less obvious examples: Light radiating from a star and striking the surface of a planet may at first sight seem to leave the star unaffected, but the momentum of each light ray hitting the planet is also imparted upon the star itself. Volition in finite intelligent agents likewise is not a causal one-way street, since the associated focusing of intention leads to a change in the cognitive content of the agent. Hence, it is probably safe to say that for all causal interactions occurring in the universe between any two substances *x* and *y*, if *x* causally affects *y*, then *y* causally affects *x*.[11]

It seems, however, that none of the reasons why causal interaction affects also the agent, rather than the patient alone, also applies to God as an absolutely unlimited being. For one, God does not occupy space, let

[11] In an interesting way, the situation is very different for *events*: if event *A* causally affects *B*, event *B* in general does not causally affect *A* (cf. Sect. 11.2). This is always true if the events have point-like extension in time, whereas it is not true of all temporally extended events or processes.

alone a finite region thereof, so that He would need to move Himself in order to causally affect an object. Rather, God is immediately present to all beings as their ultimate ontological foundation. Moreover, any causal reaction which an entity might exhibit as a result of God's action upon it could not add any positive property to an unlimited being, or augment such a positive property in it. That the interaction between an absolutely unlimited being with a limited one does not cause a change in the former is illustrated by analogy to causal interactions between greater and lesser entities in the physical universe: the greater an entity is in relation to another, the less it is causally affected through the interaction with the lesser entity, although, being finite, the greater entity is always affected to some degree as well.

13

God's Relations to Temporal Being

13.1 Are God's Relations to Creatures Real?

If God does not change as a consequence of His interaction with creatures, as I have argued in the previous chapter, it would seem—as many have objected—that the latter do not make any difference to Him, so that God cannot be said to be really related to His creatures. In Clarke's words:

> Thomistic metaphysicians seem to have been content for the most part to assert and defend the absolute immutability of God and to relegate all change and diversity to the side of the creature. But they have not gone on to explain how He can enter into a truly interpersonal dialogue with created persons, how His loving of them and their response to Him in the particular contingent ways which are proper to a free exchange between persons can truly make a difference to Him, how He is not the completely impassive, indifferent metaphysical iceberg, or at least one-way unreceptive Giver, to whom my loving or not loving, my salvation or damnation, make

D. Saudek, *Change, the Arrow of Time, and Divine Eternity in Light of Relativity Theory*, Palgrave Frontiers in Philosophy of Religion,
https://doi.org/10.1007/978-3-030-38411-1_13

no difference whatever, as Hartshorne and other process philosophers have accused Him of being.[1]

Similarly, the theologian Wolfhart Pannenberg has charged that the doctrine of divine immutability has led to the view whereby "all change in the relation of man to God cannot proceed from God, but only from man" and to Descartes' position whereby "all changes in nature must be attributed exclusively to creaturely causes, because any intervention of God in the world, once created, would be incompatible with his immutability".[2]

Part of the answer to these objections can be given in terms of what has been developed previously: If divine action does not induce a change in God, it is not necessary to attribute all change in relations between God and creatures to creatures in order to save the immutability of God. Rather, such change can also be due to divine action.

However, the objection from the lack of real relations goes still deeper, as succinctly put by William L. Craig:

> If the relation of some cause to its effect is unreal, then the cause has in particular no causal relation to its effect; that is to say, the cause is not a cause, which is self-contradictory … Thomism denies that God is literally the cause of the world, though the world is the effect of God—which seems contradictory or meaningless.[3]

Part of the difficulty here is the precise sense of what it means for a real relation to hold between two entities. In the etymological sense, a relation is real if it is "in the thing" (*res*). Another sense is that the relation holds truly, that is, objectively and mind-independently, whether or not this affects either relatum, that is, whether or not it implies some property

[1] "A new look at the immutability of God", in R. J. Roth (ed.) *God Knowable and Unknowable* (1973), here p. 45.

[2] "Die Unveränderlichkeit Gottes erzwang weiter die Konsequenz, dass alle Veränderung im Verhältnis des Menschen zu Gott nicht von Gott, sondern nur vom Menschen ausgehen kann … Ihretwegen meinte Descartes alle Veränderungen in der Natur ausschließlich auf geschöpfliche Ursachen zurückführen zu müssen, weil jeder Eingriff Gottes in die einmal geschaffene Welt unvereinbar wäre mit seiner Unveränderlichkeit." *Systematische Theologie*, vol. I (1988), p. 472.

[3] *God, Time, and Eternity* (2001), pp. 67–68.

in either relatum which is due to the relation. In Thomas Aquinas' thought, "relations which follow upon the operation of the intellect are relations of reason" whereas "as they are grounded on something which is truly in the thing … they are real relations, since they have their being grounded in the nature of the thing".[4] There is, however, an ambiguity here: is the relation grounded in things because either—or both—relata are different from what they would be if they were not so related, or rather is it due to mind-independent properties of the relata involved, and which the relata hold independently of their relation to each other? To illustrate: in the relation "*x* is a parent of *y*", the relata are not what they would be if they did not stand in this relation, but rather acquire properties because of it: parenthood, after all, affects the personality of the parent. By contrast, the relation "*x* is taller than *y*" is likewise founded on mind-independent properties of things, but neither relatum holds any property in virtue of the relation itself: Alice's being taller than Marc does not affect Marc in any way, that is, he has no intrinsic properties in virtue of this relation.

Thomas Aquinas' solution to the problem of relations between God and creatures is that this relation is real on the part of the creature, but that it is one of reason on the part of God. In this sense, it is like the relation between the knower and the known: this relation has its ontological foundation in something in the knower and is therefore real on his part, but it makes no difference in the object known, and is therefore a relation of reason on the part of the latter.[5]

I will propose, *ad experimentum*, a slightly different solution based on the distinction between the two types of real relations sketched above: the relation between God and creatures makes a difference *in re* only on the creatures' part, whereas God is not different from what He would be had He not created anything or had He created different creatures from those which He in fact created. Nevertheless, the relation of God to creatures is not a merely mind-dependent one either—as it would be if we called it a

---

[4] As cited according to L. Schütz, *Thomas-Lexikon*, 2nd edn (1895), p. 701: "relationes, quae consequuntur operationem intellectus, sunt relationes rationis, th. I. 28. 1 ob. 4; cum fundentur super aliquid, quod vere in re est … sunt reales relationes, habentes esse fundatum in natura rei, 1 sent. 33. 1. 1 c."

[5] *Summa Theologiae*, Ia, q. 13, art. 7.

relation of reason—but rather has a mind-independent *fundamentum in re*, namely that creatures ultimately owe their being to God's causal activity. In this sense, it is like the relation "greater than" between two physical objects. Propositions about such relations are true or false objectively and mind-independently, which is what we mean by the word "really" in contemporary English.[6] Thus, for example, it is mind-independently true that God is the creator of this world, but not, for example, of a possible but non-existing world containing elves and unicorns.

13.2 God's Relation to the Passage of Time and to Present, Future, and Past

Having explored the relationship between God and change in contingent beings, let's next consider God's relation to the passage of time itself. Many contemporary theistic thinkers advocate classical A-theory in order to escape the block universe in which, so to speak, everything is written, and to save the possibility of an open future, thereby allowing free choices of agents to make a difference to the history of the universe. This move, however, leads to complicated problems concerning God's relation to the passage of time and, in particular, to present, past, and future events, which according to classical A-theory differ in their ontological status.

For example, according to theologian Alan Padgett, God sustains only present events, since only these exist, but not past ones, which no longer exist. The constant change in what exists then involves God too in a corresponding change.[7] This view is endorsed by Craig:

> As God successively sustains each subsequent moment or event in being, He experiences the flow of time and acquires a growing past, as each moment elapses … So even if God is timeless sans creation, His

[6] As Clarke puts it: "Perhaps all that people finally mean when they ask whether God is 'really related' to them is whether He is truly related to them. And the answer … would be a resounding 'Yes'." Clarke (1973), p. 59.

[7] A. Padgett, *God, Eternity, and the Nature of Time* (1992), ch. 4.

> free decision to create a temporal world also constitutes a free decision on His part to enter into time and to experience the reality of tense and temporal becoming.[8]

A similar problem is Kretzmann's dilemma, already mentioned in Sect. 2.2.4, of how God can be perfect and know what time it is now. However, it must be pointed out that these problems arise on the presupposition that there is such a thing as nowness as a mind-independent property of events or world-states. The "now" is then a kind of constantly changing fact, such that for arbitrarily chosen origin and time units, there would be a definite, mind-independent matter of the fact as to what time it is now, that is, as to what events are included in the set of events occurring now.

As I have argued throughout this book, the notion of a mind-independent passage of time in the sense of a change of A-properties in things or events, and with it the notion of a moving "now", is highly implausible for a number of scientific and philosophical reasons. Craig acknowledges that the problems surrounding God's relation to the passage of time dissolve if classical A-theory is discarded and instead a tenselessly existing four-dimensional manifold is assumed.[9] We have seen, however, that a third option is available, whereby a substantival passage of time is not needed in order to escape from the changeless block. Rather, time is explained in terms of change and hence of causation, and the characteristic asymmetry of its passage from propositions concerning causal interactions between recorders and their surroundings, rather than from change in the A-properties of events. Given such a model of time, the problem of God's relation to the passage of time does not arise. God is immediately present to all things and events in spacetime as their ultimate ontological foundation, whereas the passage of time comes about only as a consequence of causal interactions between things.

Since God knows all things and all causal relations between them, it follows in particular that God knows about the relation K between any two events, that is, for any two events x and y, God knows whether Kxy

[8] *God, Time, and Eternity*, p. 60.

[9] Ibid., p. 108.

or *Kyx* or neither of these holds true.[10] Consequently, God knows all temporal relations between events as established by any frame of reference, since these temporal relations—that is, the time intervals between events as calculated by any observer in her frame of reference, equipped with a clock and knowing the spatial distances to objects surrounding her—arise from causal interactions and temporal measurement.[11] The implications of this have been elaborated in great detail by Robert J. Russell[12]: the temporal order of a sequence of events will differ according to different observers, but their respective accounts of that sequence never lead to contradictions about the invariant causal relations between these events. Russell argues that all these temporal orderings of events are present in God's eternity. The same holds for the temporal relations, as established by any frame of reference, of being past, present, or future between any two events.

If God knows the causal and temporal relations between any two events, then He also knows the world line of each endurer, together with its past and future light cones in any state of its existence. In particular, for a recorder measuring local time and associating points thereof with propositions about causal interactions between itself and its causally available surroundings, the asymmetry between a fixed past and an open future arises. This applies also to conscious observers, as argued. Thus, God knows the A-theoretical structure of spacetime associated with any state of existence of any conscious observer.

And finally, God's immediate knowledge of all things and of the causal relations between them also suffices for His knowledge of the total ordering of states of the universe—or at least the total ordering of states of the *observable* universe—established by cosmic time functions which, as we have seen, alone enable the description of the evolution of the universe as one object in such a way that the characteristics of later stages of the universe can be explained in terms of those of earlier stages.[13]

---

[10] See Sect. 3.1.2 for the definition of this relation.

[11] This is not to say that a *conscious* observer is necessarily involved in such a calculation, since temporal measurement can also be carried out by machines.

[12] *Time in Eternity* (2012), chs. 4 and 5.

[13] See Sect. 4.3.

In short, since temporal relations arise from causal interactions, and since God is present to all things and events as their ultimate ontological foundation, God does not need to be in time in order to know the temporal character of the world of creatures or to act at any point in spacetime. To posit a mind-independent A-theoretical passage of time is to unnecessarily extrapolate the perspective of us spatially limited conscious endurers, travelling through spacetime along our world lines, to the structure of reality as it is in itself. Consequently, it is equally unnecessary to assume that God is subject to the passage of time and experiences a flowing present.

This result bears upon the ancient question of whether God knows future contingents, and if so how this knowledge can be reconciled with free choice and agency. How can I be free to choose what to do tomorrow if God already knows what I will choose? This question has recently attracted much attention in the context of the debate surrounding "open theism", the movement of theologians and philosophers according to whom God does not know the future.[14] Of course, in the light of relativistic physics, it is no longer possible to non-arbitrarily define the notion of "the future of the universe" in the first place, since a global present separating past from future is no longer available. Moreover, since God is immediately present to all events, as Clarke puts it,

> The notion of any "foreknowledge," properly speaking, in God must be rejected. He does not know anything, let alone a free action, "before" it happens ... God knows all real things, past, present, and future, St. Thomas says, in their actual presence as going on, in their *now*, by His "eternal presentiality" ... Hence He knows our free actions as *contemporary with them*, not before they exist in actuality.[15]

---

[14] See, for example, C. H. Pinnock, R. Rice, J. Sanders, W. Hasker, D. Basinger, *The Openness of God: A Biblical Challenge to the Traditional Understanding of God* (1994).

[15] Clarke (1973), pp. 63–64, emphases in the original. This point was also made by Boethius: "His [God's] knowledge ... in its simple cognition beholds all things as happening now. Therefore, if you would like to consider the presence by which it discerns everything, you will more rightly estimate it, not as foreknowledge of, so to speak, 'the future' but as knowledge of a never ceasing instant."

In the original: "scientia eius ... omnia quasi iam gerantur in sua simplici cognitione considerat. Itaque si praesentiam pensare velis qua cuncta dinoscit, non esse praescientiam quasi futuri sed scientiam numquam deficientis instantiae rectius aestimabis." *Consolatio philosophiae* V, P6, 60–68.

Even in pre-relativistic times, when the notion of a global present was by and large taken for granted, it was realized that the idea that the knowledge of an event—whether present or future—would impose a necessity on that event is due to a confusion between *necessitas consequentiae* and *necessitas consequentis*,[16] a point which was discussed in Sect. 3.3 in the context of the Rietdijk-Putnam argument from special relativity to universal temporal determinism.

All this is not to say that there are no philosophical problems concerning the relationship between God's immediate sustaining of our mental activity and the possibility of free choice.[17] But this is an issue of causality, not temporality. The problem of how God's knowledge of contingent events "before" they occur can be reconciled with human freedom simply dissolves, since it arises from a projection of our A-theoretical experience of time onto God.

Finally, the "third option" in philosophy of time developed in this book can be applied to God's relationship to the irretrievable and unchangeable past. As Aristotle put it,

> Nothing that is past is an object of choice, e.g. no one chooses to have sacked Troy; for no one deliberates about the past, but about what is future and capable of being otherwise, while what is past is not capable of not having taken place; hence Agathon is right in saying "For this alone is lacking even to God, to make undone things that have once been done".[18]

In Chap. 9, the fixity of the past and the openness of the future were reconstructed in terms of a changing and time measuring recorder A and the assignment of sets P_t to local times t. Once this assignment has been made, the principle of non-contradiction itself excludes that t and A_t are associated with a set P_t' different from P_t. The same does not apply to A's

[16] Cf. Boethius *op. cit.* V, P. 4, 56–60, and V, P. 6, 91–102. In the latter passage, Boethius distinguishes between the "necessitas simplex" and the "necessitas condicionis", a distinction which later became known as that between the *necessitas consequentis* and the *necessitas consequentiae.*

[17] This problem would need to be discussed in the context of the much larger issue of contingency and the principle of sufficient reason, which is beyond the scope of this book. However, for some in my view insightful reflections on free choice and divine causality in conscious agents, cf. Clarke (1973), pp. 67–70.

[18] *Ethica Nicomachea*, 1139b, 6–11.

future, since no definite set of propositions is assigned to future stages of *A*. The tensed character of time as experienced by an agent is sufficiently explained through the successive assignment of such sets of propositions. It does not require a tensed character of the world itself.

By the above, while God can act anywhere where He is causally present, that is, throughout spacetime, the principle of non-contradiction excludes His acting to change *A*'s past. For such divine action would mean that a set P_t' is assigned to a time t to which a set P_t (different from P_t') is assigned. That it is not possible that God change the past is therefore not due to some mind-independently existing temporal determinism, which would somehow fix past events and leave future ones open, but rather to the more basic fact that the principle of non-contradiction cannot be violated.

14

The Unlimited, the Limited, and the World

The drift of this book can be summarized by reading it backwards, starting with God as the unlimited ground of being: since God cannot exist in two states (as argued in Sect. 12.2), He cannot be subject to change. But this does not mean that, conversely, any limited being necessarily changes, since it is quite possible for there to be a limited being having only one actualized state. Limitedness is therefore necessary but not sufficient for change. Once, however, a limited substance has at least two states, and in addition there is an object able to record these states, we obtain a before-order between them and therefore find ourselves in a situation of change which is at least what I have called "*aevum*-like". Next, events fall into types (such as "sunset" or such like) of which there are several tokens. Records of different tokens of a particular type of event—and these tokens, just as their corresponding records, may even be intrinsically indiscernible from each other—are, as we have seen, sufficient for establishing a local time count. Finally, as has been shown, this local time has the familiar arrow-like structure.

It is illustrative to compare this general picture of the relationship between eternity, change, and time with Plato's as given in the *Timaeus*:

D. Saudek, *Change, the Arrow of Time, and Divine Eternity in Light of Relativity Theory*, Palgrave Frontiers in Philosophy of Religion,
https://doi.org/10.1007/978-3-030-38411-1_14

> When the father creator saw the creature which He had made moving and living, the created image of the eternal gods, He rejoiced, and in His joy determined to make the copy still more like the original; and as this was eternal, He sought to make the universe eternal, so far as might be. Now the nature of the ideal being was everlasting, but to bestow this attribute in its fullness upon a creature was impossible. Wherefore He resolved to have a moving image of eternity, and when He set in order the heaven, He made this image eternal but moving according to number, while eternity itself rests in unity; and this image we call time. For there were no days and nights and months and years before the heaven was created, but when He constructed the heaven He created them also. They are all parts of time, and the past and future are created species of time, which we unconsciously but wrongly transfer to the eternal essence.[1]

Plato thus draws an intimate connection between time and the impossibility of creation imitating eternity perfectly. Loosely, we might say that what is created, in contrast to eternity, is not "all at once" and in this way is capable of changing, a general picture in consonance with the one painted in this book. The main difference seems to be that, in Plato's view, it is the imperfect nature of the world as a whole which brings it into time. This time in turn, in accordance with the ancient worldview, was provided by the eternal spheres' motions around the earth, distinguished from each other only by number.[2] According to my account, developed before the background of the absence of a global time, about which Plato of course could not know, it is the limitedness of substances which, via the steps briefly recapitulated above, gives rise to time. Proceeding in this way leads directly to a substance-centred, and hence local, account of time. The common denominator, however, is that temporality must be understood as arising from imperfect, partial realization of being, in contrast to the fullness of being.

Does this mean that there is only divine eternity on the one hand, and isolated objects and happenings on the other, with no such thing as a world holding them together? In the absence of a global time, it is true that we cannot think of the world, naïvely, as an extent of space containing

[1] *Timaeus*, 37 c–e.

[2] Cf. the discussion in T. Roark, *Aristotle on Time* (2011), pp. 22–33.

everything, in the same way in which a room contains furniture, because the notion of an extent of space is intimately connected to that of simultaneity—when we conceive of such an extent, we picture a set of points existing simultaneously.[3] A better way of thinking of the world, I suggest, is as the set of all possibilities actualized, perhaps together with the relations between them, or alternatively, with Wittgenstein, as the "totality of facts".[4] Similar considerations, incidentally, apply to multiverse cosmological models of the type where one bubble universe spawns another before the background of an underlying time arrow which is simply presupposed. As is now sufficiently clear, such a presupposition, once again, rests on a pre-relativistic extrapolation of time as experienced locally to reality as a whole.[5] Pointing out this conceptual problem with such cosmological models does not, of course, mean attempting to limit what exists to what we conceive of as our universe. But it does mean that, departing from a naïve view of the world as a container set in absolute time, and instead adopting a notion of "world" as suggested above, whatever possibilities are actualized are *ipso facto* part of the world, however these actualizations may be spatiotemporally related to us.

Despite the above, we are dealing with more than fragmented, unconnected things and happenings. For things interact causally, and causal interactions in spacetime lead to records contained in recorders, from which true statements can be obtained. If we consider the set of all such statements, obtainable from any records in spacetime, then any combination of these statements, given that they are true, must be consistent. This does not mean agreement between all observers on every point, in particular on values of quantities such as length, duration, or mass, but it does mean invariance of causal stories—the stories of what happened and what interactions there were in spacetime. The consistency of the world was important to Einstein, and the finiteness of the speed of light guaranteed it by guaranteeing the invariance of timelike-related events. Similarly, we have seen, in Sect. 11.2, that this requirement of consistency,

[3] Cf. the observations in W. Pannenberg, *Systematische Theologie*, vol. II (1991), chapter 7, section II.2.c.

[4] *Tractatus logico-philosophicus*, 1.1.

[5] Cf. S. E. Rugh, T. Zinkernagel, "A critical note on time in the multiverse" (2013).

together with probabilistic considerations about causal interactions, leads to the invariance of the relation of causal connectability. As far as I can see, said requirement—which neither needs nor is capable of further proof—does not disqualify particular matter and energy distributions as causally pathological, and therefore absurd. For example, for reasons given in Sect. 11.3, it does not rule out a Gödel universe, although it does rule out backwards time travel. The consistency of causal stories, then, can be thought of metaphorically as the "glue" which makes us able to think of the world as an integrated whole, a glue which space cannot provide. It is, I think, profoundly linked to the nature of God, who does not contradict Himself and whose creative "yes" is never a "no".

Open questions remain, of course. One of particular relevance for the topic of this book is how the asymmetric notions of causation and causal dependence relate to the asymmetric arrow of time. I suspect that both are linked, in different ways, to the timeless asymmetry of inclusion which features in logic and set theory, and which was introduced in Sect. 9.1.1. But spelling out these connections is the subject of future work. We can make headway by solving problems in our understanding of the world, which after all is the task of philosophy, bit by bit and without having answers to everything. As for this book, it is my hope that it will contribute to a better understanding of the arrow of time, and thereby also of the relation between temporal and eternal being.

Appendix A: Outline of Special Relativity

A key achievement of nineteenth-century physics has been the unification of electric and magnetic phenomena embodied in Maxwell's equations, formulated by the early 1860s after decades of experimental and theoretical work in the area. When these equations are solved in the absence of electrical charges and currents, a solution is obtained which can be interpreted as an electromagnetic, periodic disturbance—that is, an electromagnetic wave—travelling through space. It is found that the speed of such a wave must equal a constant equal to about $3 \cdot 10^8$ metres per second, a value now known as c.[1] Since light was known to travel at this speed, Maxwell surmised that light itself is an electromagnetic wave. But relative to what do electromagnetic waves travel at speed c? Maxwell's equations themselves gave no indication, but it was natural, in the context of nineteenth-century physics, to assume that c was the speed of light relative to the medium it was travelling in, that is, the ether which was assumed to fill empty space. As is well known, the attempts by Michelson and Morley to measure an ether-wind due to the motion of the earth rela-

[1] See, for example, H. D. Young, R. A. Freedman, *University Physics*, 13th edn (2012), pp. 1055–1057.

D. Saudek, *Change, the Arrow of Time, and Divine Eternity in Light of Relativity Theory*,
Palgrave Frontiers in Philosophy of Religion,
https://doi.org/10.1007/978-3-030-38411-1

tive to the ether failed. Thus, by the end of the century, the possibility that light might travel through nothing but empty space, while philosophically puzzling, had to be taken seriously.

In addition, centuries of physics had corroborated the principle known as Galilean (sometimes also called Newtonian) relativity: the physical laws which one will verify when situated in a frame in uniform motion are indistinguishable from those one will verify in a frame at rest. Thus, either frame can with equal right be called the "rest frame". A frame in uniform motion, or an inertial frame (as opposed to an accelerated one), is one travelling at a constant velocity relative to a frame considered to be at rest.[2]

Combining what has been said in the last two paragraphs, it follows that Maxwell's equations will hold true in any inertial frame and that, therefore, light will travel at speed c in any such frame. Thus, for example, an observer standing on earth will measure the incoming sunlight to have speed c, and an observer travelling towards the sun at great speed will obtain the same result. Contrary to what we would expect from the Galilean theorem of composition of velocities, which agrees with everyday experience, the velocity of the "moving" observer is not added to c. Nor is it possible, by the same reasoning, to catch up with a light ray and travel alongside it or overtake it. Thus, not only is c fixed, but it also sets an absolute speed limit on all motion.

The fixity of c leads to relativity of simultaneity. This is best understood using the following classic thought experiment: A train is in motion relative to the surface of the earth. An observer A ("she") is standing on the platform, while another, B ("he"), is sitting at the midpoint between the ends of the train. A bolt of lightning strikes each end of the train, leaving marks on the ground below. The light signals[3] reach A simultaneously. She then notices that she is situated half-way between the marks on the ground. Since the speed of light is independent of the source which emits it, she reasons that either light signal must have taken a time equal to half the length of the train, divided by the speed of light, to reach her. She concludes that the two bolts must have struck the end of the train simultane-

[2] On the philosophical problem of identifying accelerated frames versus inertial ones, see B. Dainton, *Time and Space*, 2nd edn (2010), ch. 11.

[3] In the following discussion, we will always talk of light signals, although of course any electromagnetic signals can be used.

ously. On the other hand, *B*—as viewed from the surface of the earth—is travelling towards the light signal coming from the front of the train, and away from the light signal coming from the back. Therefore, the signal from the front will reach him first. But, again due to the fixity of *c*, he too knows that the time each signal has taken to reach him is $L/2c$, where L is the length of the train.[4] He concludes that lightning must have struck the front end of the train before the back end, in disagreement with *A*'s result.

Note that "simultaneous for an observer" in this context does not mean "experienced simultaneously by an observer". Rather, our two observers have reached their conclusions about simultaneity or non-simultaneity based on calculations involving lengths and the speed of light.

Relativity of simultaneity in turn results in relativity of length measurements. The reason is very simple: in order to measure the length of a moving object, the distance of either end of the object from a chosen origin must be measured simultaneously. Hence, if two observers in different states of motion disagree about simultaneity, they will also disagree about the lengths of objects. Consider again the marks left on the railway track in the previous thought experiment: According to *A*, both lightning bolts struck the train simultaneously, so that the distance between the marks must equal the length of the train. But according to *B*, the front end of the train was struck first. Then, the train moved forward, whereupon the second bolt struck. Hence, the distance between the marks on the track is shorter than the length of the train according to *B*. By this reasoning, an observer in motion relative to an object will measure its length along the direction of motion to be shorter than the length as measured by an observer at rest relative to the object. Thus, the length measured in the object's own rest frame will be longest. It is called the proper length.

Next, suppose that two objects are in relative motion along the x-axis, but that there is no motion along the y-axis. There will then be relativistic length-contraction along the x-axis, but not along the y-axis, so that both observers will be in agreement about lengths in the y-direction.[5]

---

[4] It does not matter, for the sake of this thought experiment, that *A* and *B* will disagree on the length of the train. What is relevant here is that both observers conclude that the travel time for either light signal is the same.

[5] Indeed, the assumption of relativistic changes of lengths along the y-axis leads to contradictions, as shown in E. F. Taylor, J. A. Wheeler, *Spacetime Physics: Introduction to Special Relativity*, 2nd edn (1992), pp. 65–67.

All of the above then entails relativistic stretching of temporal measurements, known as time dilation. Here is why: Time is measured by counting oscillations of regularly oscillating systems. We are familiar with mechanical clocks, but since mechanical interactions are mediated by the electromagnetic force, it is simplest, for the derivation of time dilation, to think of a light clock, consisting only of a light ray bouncing back and forth between two mirrors. Next, suppose again that two observers are in relative uniform motion along the *x*-axis, but not along the *y*-axis. If they wish to compare their time-measurements, the axis of the light clock must be placed in the *y*-direction, since placing it in the *x*-direction would lead to disagreement about the length of the clock's axis.

The observer at rest relative to the clock (let's again call him *B*) will say that half a period of one oscillation is equal to L/c, where L is the length of the clock's axis. But for observer *A*, *B* and his clock are moving at speed v. Thus, a light signal fired from the bottom of the clock will need to travel a distance $\sqrt{L^2 + v^2 t'^2}$ before it hits the top, where t' is the time between emission and reception as measured by *A*, as illustrated below. The travel time is $\frac{\sqrt{L^2 + v^2 t'^2}}{c}$. Thus, a moving clock ticks slower than one at rest.

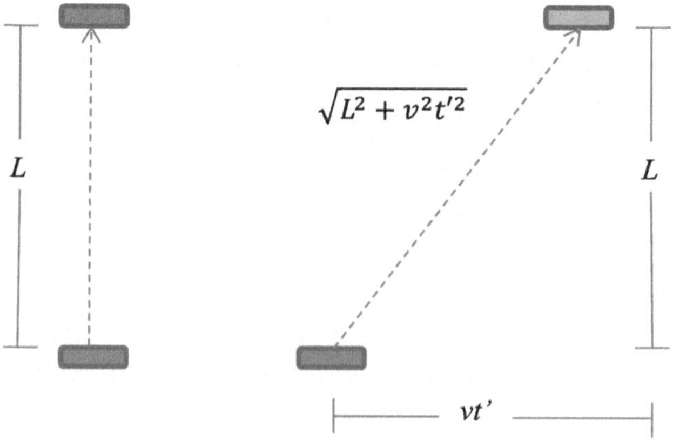

Fig. A.1 A light clock in two frames of reference

Therefore, observers will disagree about lengths lying in the direction of relative motion (which we here call the *x*-direction), and about time intervals, but not about lengths perpendicular to the direction of motion. But, as can be seen from the graph, for any relative velocity between the two observers, the time interval is determined by the length of the hypotenuse of the triangle. Let's use this distance to measure time and call it t'.[6] Then, the quantity $t'^2 - x'^2$ will be the same according to all observers in uniform relative motion, and for any two reference frames *F* and *F'* associated with such observers, we get

$$t^2 - x^2 = t'^2 - x'^2 \equiv \Delta s^2.$$

In this context, time can be measured in spatial units (e.g. metres), as we did above. Alternatively, it can be measured in temporal units (e.g. seconds), in which case, however, spatial distances must also be measured in temporal units, which is done by dividing them by the speed of light. The latter choice is typically used in astronomy, where distances are often given in light years.

The quantity Δs^2 is called the invariant spacetime interval.[7] Given it, spatiotemporal separations between events naturally fall into three types: Where the time-component (t) is greater than the spatial one (x), Δs^2 is greater than zero, and thus for any reference frame, we must have $t'^2 > x'^2$. We then speak of a "timelike" interval. Analogously, where the spatial component dominates, we have $\Delta s^2 < 0$. Such an interval is called "spacelike". Where $t = x$, the spacetime interval is denoted as "lightlike". Consider, for example, an event organized by hypothetical inhabitants of planet Pluto, which is about five light-hours away. If the event takes place at a time greater than or equal to five hours from the time when you are reading these lines, you could send a light signal to Pluto in order to influence it. The spacetime interval between the event on Pluto and your

---

[6] That is, the speed of light is, by convention, set equal to 1, so that the distance ct is replaced by t.

[7] A quantity is called invariant if it is the same according to all frames of reference. In this context, we are of course talking only of inertial frames.

reading these lines is then timelike or lightlike. But if the event takes place sooner, then it is spacelike-separated from you, and you can do nothing about it.

If $t^2 - x^2 > 0$ for a pair of events in some frame F, we can always find another frame F' such that $x' = 0$, since the two events can always be connected by a signal slower than light. In other words, F' is the frame travelling at a speed less than c relative to F, such that the two events are situated in the same place. Likewise, if $t^2 - x^2 < 0$, it is always possible to find a frame F' such that $t' = 0$.

In either case, the inequality is obeyed if t is replaced by $-t$. Does this mean that the time order of events is reversible? For spacelike-separated events, there will always be a frame of reference such that their order of occurrence is reversed. However, for timelike-separated events, no such frame can be found, since it would need to travel faster than the speed of light. Indeed, switching the order of such events leads to a *philosophical* difficulty: For suppose that A and B are two timelike-separated events, that is, according to some observer, $t = t_B - t_A > 0$. Then, according to this observer, A can influence B. In particular, A can bring about B. If there were a frame of reference such that the sign of the time separation were switched, B would happen before A, and B could influence A according to an observer in this frame. Then, it would also be possible for B to prevent A. But A was assumed to cause B, so that B, by preventing A, could prevent its own occurrence. The assumption that the sign of the time component could be switched leads to a universe in which events could prevent themselves, and the fact that no frame of reference can travel faster than c excludes such a possibility.[8]

Given all this, the spacetime surrounding any event e (situated at the origin) is divided into three regions, as illustrated:

[8] In fact, the philosophical argument presented here relies on the simplifying assumption that causes precede their effects, an assumption which this book challenges. For this reason, I give an alternative account of the invariant temporal asymmetry of timelike-separated events in Sect. 11.2.

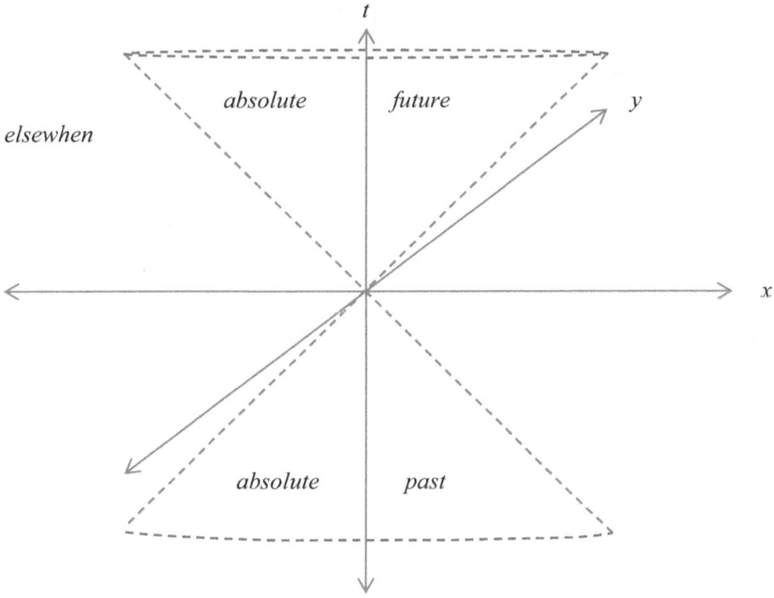

Fig. A.2 The light cone structure of the spacetime around an event

Those events separated by a spacetime interval $\Delta s^2 \geq 0$ fall into two regions: into the "absolute future" if their time-coordinate is positive and into the "absolute past" if it is negative. The former events can be influenced from the spacetime location of *e* through a light signal or, if $\Delta s^2 > 0$, by a signal slower than light. Spacelike-separated events are situated in the region surrounding the light cone, known as the "elsewhere" or "elsewhen".

Note that, for the sake of illustration, one spatial axis was suppressed. In fact, the volume of the cone is four-dimensional.

It can be shown[9] that the spatial and temporal measurements carried out by two observers situated in inertial frames F and F' are related by

$$t' = \frac{-\beta x + t}{\sqrt{1-\beta^2}}$$

[9] See Taylor, Wheeler (1992), pp. 99–103.

$$x' = \frac{x - \beta t}{\sqrt{1-\beta^2}}$$

$$y' = y,\ z' = z,$$

where the x-direction is assumed to be the direction of relative motion, the speed of light is set equal to one, x and t are measured in the same units and β is the fraction of the speed of light at which F and F' are moving relative to each other. These equations are known as the Lorentz transformations.

Appendix B: Sketch of General Relativity

Only a rough outline can be given here of a subject which is among the most difficult in physics. I will therefore focus on the key ideas, referring the reader to sources containing precise derivations on the way.[10]

From the Equivalence Principle to Spacetime Curvature

The special theory of relativity (SR) on its own is inadequate for understanding the structure of spacetime, since it cannot accommodate the most important long-range interaction between physical bodies, that is, gravity:

[10] For relatively accessible introductions to General Relativity, see especially R. Mould, *Basic Relativity* (1994); B. F. Schutz, *A First Course in General Relativity*, 2nd edn (2009); R. D'Inverno, *Introducing Einstein's Relativity* (1992); and T. P. Cheng, *Relativity, Gravitation and Cosmology*, 2nd edn (2010); and in my view very clear exposition of the indispensable subject of tensor calculus and differential geometry is P. Grinfeld's *Introduction to Tensor Analysis and the Calculus of Moving Surfaces* (2013). Basics of tensor calculus are also well explained in L. R. Lieber, *The Einstein Theory of Relativity* (2008); and D. Fleisch, *A Student's Guide to Vectors and Tensors* (2012), ch. 4–6. A very useful, more qualitative and mostly non-technical source on these topics is found in G. F. R. Ellis, R. M. Williams, *Flat and Curved Space-Times* (2000).

D. Saudek, *Change, the Arrow of Time, and Divine Eternity in Light of Relativity Theory*, Palgrave Frontiers in Philosophy of Religion,
https://doi.org/10.1007/978-3-030-38411-1

Newton's inverse-square law of gravitation cannot be formulated in a frame-independent way given that the distance between bodies depends on the frame of reference. Furthermore, the classical understanding of gravity had to be modified for two principal reasons: First, in Newton's equations, the gravitational interaction is instantaneous, depending only on distance, but not on time—in contradiction to the basic postulate of SR whereby no influence can travel at a speed faster than c. Second, Newtonian gravity could not explain the fact that the gravitational mass of a body m_G is always found to be equal to its inertial mass m_I. The former is the "gravitational charge" of a body entering into Newton's law:

$$F_G = \frac{GMm_G}{r^2},$$

where M is assumed to be the mass of a large body exerting gravitational pull onto m_G. The acceleration of m_G in response to this force is F_G/m_I. But since $m_I = m_G$, the acceleration of any body in the vicinity of M is independent of its mass, leading to the familiar result that all bodies fall at an equal rate given by GM/r^2.

Pondering about gravity, Einstein arrived at a key insight, which he called "the happiest thought of my life" and which is now known as the equivalence principle: motion under the action of gravity is locally indistinguishable from inertial motion. Consider a person in a room in free fall towards the earth: if she throws objects through the room, these will fall towards the earth at the same rate as she and the room do, so that she will see these objects to travel in a straight line. The physics in such a room is just like that which holds in empty space far away from the influence of gravitating bodies. The only difference between these two frames arises because of tidal effects, that is, because of the varying strength of the gravitational field: Objects close to the bottom of the room will accelerate towards the earth faster than those at the top, so that distances between such objects will gradually increase. Also, objects at the left and right end of the room will both fall towards the centre of gravity of the earth, so that distances between them will decrease.

In classical mechanics, objects on which no force acts travel in a straight line, which is the line of least distance between any two points on an object's path. In the same way, in GR an object follows a geodesic, which is the line of least spacetime interval between any two spacetime points on its path.

If the path in spacetime of a falling object is an inertial one, that of an object at rest in a gravitational field is an accelerated one. Such objects undergo forced motion in spacetime and can be viewed as "accelerating upwards". A rocket accelerated at 9.8 m/sec^2 in empty space is indistinguishable from a rocket at rest on the surface of the earth—again, apart from tidal effects.

This, in turn, leads to relativistic time dilation. Consider a rocket undergoing constant acceleration, with clocks fixed at the "top" and "bottom" ends. The clocks are of the same build and emit a monochromatic light ray every second. After a photon is emitted from the bottom end, the rocket picks up speed, so that the photon is received at a reduced frequency at the top end and is therefore red-shifted. An observer at the top of the rocket will conclude that the frequencies received from the bottom are "too low". But since the clocks are of the same build and emit the same monochromatic light, the only way in which this can happen—from the point of view of an observer at the "top" end of the rocket—is that time passes more slowly at its "bottom" end. Exactly the same applies to light rays fired off from clocks at rest situated at different gravitational potentials. Put differently, an upward-travelling photon cannot slow down, as a material particle would, due to the fixity of c. But it must lose energy, which is only possible through a reduction of its frequency, because of the Einstein-Planck relation whereby the energy of a photon is proportional to its frequency ($E = h\nu$). Thus, time passes more slowly at a lower gravitational potential than at a higher one. The relationship between time units τ_1 and τ_2 of two observers situated at gravitational potentials Φ_1 and Φ_2 is given by[11]

$$\tau_1 = \left(1 + \frac{\Phi_1 - \Phi_2}{c^2}\right)\tau_2.$$

Spatial measurements may likewise be affected by the presence of matter, and it turns out that in fact they are. Hence, basis vectors of a spacetime diagram change as a function of position. We have already seen that

[11] Cheng (2010), p. 72.

they do so for inertial frames in relative motion (see the illustration in Sect. 3.1.1), but in the current context it is the presence of matter which causes a coordinate system to curve. General relativity (GR) therefore had to be formulated in the mathematics of changing coordinate systems: differential geometry and tensor calculus.

A Bit of Tensor Calculus

Consider a position vector sitting in a coordinate system. If the underlying coordinate system changes, but the vector does not, the vector's components will change. For example, if the coordinate system C is rotated by a certain angle α, the components of the vector in the rotated coordinate system C' will be the same as the components of a vector rotated by angle $(-\alpha)$ in C. Intuitively, vector components change in the opposite sense than do the basis vectors, which is why vector components are known as "contravariant" quantities, whereas basis vectors are "covariant". Of course, spacetime will in general twist coordinate systems in much more complicated ways. It is found that covariant and contravariant quantities combine to yield invariants. For example, the same vector $\boldsymbol{v}$ (in a three-dimensional space) can be written in C and in C' as

$$\boldsymbol{v} = v_1 \boldsymbol{e}_1 + v_2 \boldsymbol{e}_2 + v_3 \boldsymbol{e}_3 = v_1' \boldsymbol{e}_1' + v_2' \boldsymbol{e}_2' + v_3' \boldsymbol{e}_3',$$

where $\boldsymbol{e}_i$ is the i-th basis vector.

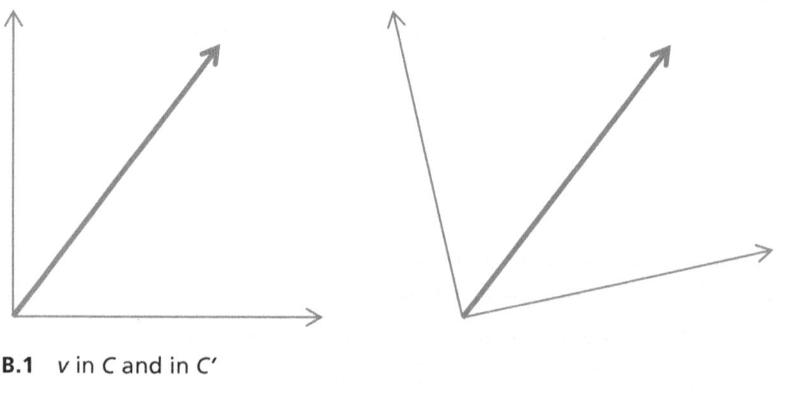

Fig. B.1 v in C and in C'

If we assign upper indices (not to be confused with exponents!) to contravariant quantities, and lower indices to covariant ones, then a vector can be written as $\sum_i v^i \boldsymbol{e}_i$. Einstein introduced the convention of dropping the summation sign whenever the same index appears twice, that is, once as a covariant and once as a contravariant index, so that we can simply write $v^i \boldsymbol{e}_i$ to represent a vector.

An important scalar invariant is the length of a vector. It can be shown[12] that the square of the length of a differential vectorial displacement in a curved coordinate system is given by

$$ds^2 = \sum_{i,j} g_{ij} dx^i dx^j \equiv g_{ij} dx^i dx^j .$$

Here, g_{ij} is the dot product of the i-th and j-th basis vectors. It is known as the metric Tensor and can be written as an $n \times n$ matrix containing all of these dot products, where n is the dimension of the space. Since $\boldsymbol{e}_i \bullet \boldsymbol{e}_j = \boldsymbol{e}_j \bullet \boldsymbol{e}_i$, the metric Tensor is symmetric. In relativity, we have four axes. Traditionally, the time coordinate of an event is called x^0, and the spatial ones x^1, x^2 and x^3. Here, the quantity ds^2 does not refer to spatial length, but to the invariant spacetime interval between events. Also, the indices in four-dimensional space are denoted by Greek letters, so that instead of the above we write

$$ds^2 = g_{\mu\nu} dx^\mu dx^\nu .$$

Consider again our vector $\boldsymbol{v}$ in a coordinate system whose origin remains fixed. Its primed components in C' can then be computed from its unprimed components in C as follows:

$$V'^\mu = \sum_\nu \frac{\partial x'_\mu}{\partial x_\nu} V^\nu \equiv \frac{\partial x'_\mu}{\partial x_\nu} V^\nu ,$$

where again the summation sign is dropped, being implied by the repeated indices in accordance with the Einstein convention. That is, the

---

[12] See, for example, Fleisch (2012), pp. 140–144.

primed components of $\boldsymbol{v}$ in C' are a linear combination of all of its components in C, where the coefficients are partial derivatives of the variable $x^{\mu'}$ with respect to each variable x^{ν} in C.[13] The components of $\boldsymbol{v}$ are an example of a tensor—more precisely, a contravariant tensor of rank one. By definition, a rank one contravariant tensor is one for which the relationship just given holds true. Similarly, a rank two contravariant tensor is an object whose components in primed and unprimed coordinate systems are related by

$$V'^{\mu\nu} = \frac{\partial x'_{\mu}}{\partial x_{\sigma}} \frac{\partial x'_{\nu}}{\partial x_{\tau}} V^{\sigma\tau}.$$

For example, the tensor product of two contravariant tensors of rank one (V and U) is a contravariant tensor of rank two. That is, if each component V^{μ} is multiplied with each U^{ν}, we get a new tensor W for which the relation

$$W'^{\mu\nu} \equiv V'^{\mu} U'^{\nu} = \frac{\partial x'_{\mu}}{\partial x_{\sigma}} \frac{\partial x'_{\nu}}{\partial x_{\tau}} V^{\sigma} U^{\tau} \equiv \frac{\partial x'_{\mu}}{\partial x_{\sigma}} \frac{\partial x'_{\nu}}{\partial x_{\tau}} W^{\sigma\tau}$$

holds. There are also covariant tensors. For example, it can be shown that the metric Tensor $g_{\mu\nu}$ is a covariant tensor.[14] It has rank two, and just like with any covariant tensor of rank two, its primed components are related to its unprimed ones by the relationship

$$g'_{\mu\nu} = \frac{\partial x_{\sigma}}{\partial x'_{\mu}} \frac{\partial x_{\tau}}{\partial x'_{\nu}} g_{\sigma\tau}.$$

All this may be a little confusing, but the important point is this: From the above transformation equations, if all components of a tensor in C are zero, they will also be zero in C'. Also, the sum or difference of two tensors is itself a tensor. From these two facts it follows that, if a tensor equa-

[13] See Lieber (2008), pp. 127–159, for some illustrative examples.

[14] Grinfeld (2013), p. 81.

tion is true in one coordinate system, then it is true in any other coordinate system. For if A and B are tensors, then

$$A = B \Leftrightarrow A - B = 0. \text{ Then also } (A - B)' = A' - B' = 0 \Leftrightarrow A' = B'.$$

It bears repeating: a tensor equation holds true independently of the frame of reference in which the individual tensors which enter into the equation are computed. It is chiefly for this reason that tensors are said to express "the facts of the universe".[15]

Covariant Derivative

A concept of key importance in differential geometry is that of the *covariant derivative*: Consider a vector field on a curved coordinate system, and suppose that we wish to find the partial derivative of it with respect to the x^μ-axis. If the coordinate system were not curved, we would find this derivative by moving an infinitesimal displacement along that axis from x^μ to $(x^\mu + \delta x^\mu)$, subtracting the vector Q sitting at the second point from the vector P at the first point and then dividing by the distance δx^μ. But of course, in a curved coordinate system, this will yield the wrong result, since the components of the vector Q are calculated with respect to different basis vectors than those of the vector at P. To find the "absolute" derivative[16] of the ν-component of the vector V with respect to x^μ, we must therefore first find its partial derivative with respect to x^μ and then add a correction term to it:

$$(\text{absolute derivative}) = (\text{partial derivative}) + (\text{correction term})$$

By definition, the precise relationship is[17]:

[15] Fleisch (2012), p. 97.

[16] The discoverers of covariant differentiation, G. Ricci-Curbastro and T. Levi-Civita fittingly denoted the technique by the name of "absolute differential calculus". See Grinfeld (2013), p. 105.

[17] Grinfeld (2013), p. 108.

$$\nabla_\mu V^\nu = \frac{\partial V^\nu}{\partial x^\mu} + \Gamma^\nu_{\mu\sigma} V^\sigma,$$

where the symbol ∇ denotes covariant differentiation. Note that the last term is in fact a sum of four terms (assuming that we are in four-dimensional space). The symbol $\Gamma^\nu_{\mu\sigma}$ is called a Christoffel symbol. It is defined by[18]

$$\frac{\partial \boldsymbol{x}_\mu}{\partial x^\nu} = \Gamma^\sigma_{\mu\nu} \boldsymbol{x}_\sigma.$$

That is, the partial derivative of the μ-basis vector with respect to the ν-axis is a linear combination of all of the basis vectors. The Christoffel symbol $\Gamma^\sigma_{\mu\nu} \boldsymbol{x}_\sigma$ is then the coefficient multiplying the σ basis vector.

Covariant differentiation can also be understood through the idea of parallel transport: The "absolute" derivative between the two vectors *P* and *Q* can be found by transporting the vector *P* to the point where *Q* is. Then, the components of *P* and *Q* are calculated with respect to the same basis.

Intrinsic Curvature of a Space

Parallel transport can be used to reveal and quantify the intrinsic curvature[19] of a space. On a curved space, transporting a vector first in the μ-direction and then in the ν-direction will yield a different result than transporting it first in the ν-direction and then in the μ-direction. In a

---

[18] Grinfeld (2013), p. 67.

[19] Intrinsic curvature is to be distinguished from extrinsic curvature. A surface is intrinsically curved if parallel lines on it will converge or diverge. Extrinsically curved surfaces are curved only by reference to a space they are embedded in, but parallel lines on such a surface will not converge or diverge. The classic example of intrinsic curvature is the sphere: if two people situated at different places on the equator walk northwards, parallel to each other, they will meet at the north pole. By contrast, the surface of a cylinder is curved only relative to the surrounding space. Parallel lines drawn on it will remain apart. See especially Schutz (2009), pp. 163–173.

flat space, both operations yield the same result. Mathematically, this means that, for a curved space,

$$\nabla_\mu \nabla_\nu V^\sigma \neq \nabla_\nu \nabla_\mu V^\sigma \Leftrightarrow \left(\nabla_\mu \nabla_\nu - \nabla_\nu \nabla_\mu\right) V^\sigma \neq 0.$$

Computing the operator in brackets, called a commutator, for combinations of axes μ, ν in a space then results in a rank four tensor which gives complete information about the curvature of the space at any point. This tensor is called the Riemann-Christoffel Tensor, denoted as $R^{\sigma}_{.\tau\mu\nu}$. Explicit computation shows that it is given entirely in terms of Christoffel symbols and derivatives of Christoffel symbols.[20]

From the Riemann-Christoffel Tensor, it is possible to derive a rank two tensor called the Ricci tensor $R_{\mu\nu}$, from which in turn a scalar R can be computed.[21] All of these mathematical objects serve as a diagnostic of the intrinsic curvature of a space.

The Einstein Field Equation

Since the distribution of matter and energy determines the curvature of spacetime, there ought to be a tensor equation relating the two. The former is captured in a tensor known as the stress-energy tensor $T^{\mu\nu}$.[22] For the purposes of cosmology, this tensor is found by assuming that the large-scale distribution of matter and energy in the universe can be modelled by that of an ideal gas. In other words, the universe is viewed as a homogenous soup made of galaxies. It is not obvious that this assumption is justified since, at an intermediate scale, matter is distributed neither homogenously nor isotropically. Rather, there are dense regions containing galaxies and clusters thereof, interspersed with large voids. But at very large scales, these inhomogeneities level out, and the distribution of matter in the universe can be viewed as homogenous and isotropic. $T^{\mu\nu}$ is defined by the requirement that its rate of change—that is, its

[20] See, for example, Lieber (2008), pp. 200–206; Fleisch (2012), pp. 183–192.

[21] See, for example, Mould (1994), pp. 307–308.

[22] See Mould (1994), ch. 11, for a derivation of this tensor.

covariant derivative—be equal to the external force density, or pressure, applied. If there is no net force, as is the case at any point in a fluid at rest, the covariant derivative must be zero.

For this reason, the covariant derivative of the tensor describing space-time curvature must also be zero. In addition, the sought-for tensor must be of rank two, like the stress-energy tensor. In an astonishing mathematical tour de force, Einstein was able to show that both of these requirements are met by a linear combination of the Ricci Tensor and the metric Tensor, now known as the Einstein Tensor.[23] In this way, he arrived at a relation which in covariant form reads

$$R_{\mu\nu} - \frac{1}{2} R g_{\mu\nu} = -k T_{\mu\nu}.$$

The constant of proportionality k is calculated to be equal to $8\pi G/c^4$, where G is the Newtonian universal gravitational constant. Thus, the Einstein field equation is

$$R_{\mu\nu} - \frac{1}{2} R g_{\mu\nu} = -\frac{8\pi G}{c^4} T_{\mu\nu}.$$

The requirement that the covariant derivative of the tensor on the left-hand side be equal to zero is also met when the metric Tensor times a constant is added to or subtracted from it, since the covariant derivative of the metric Tensor is by definition zero.[24]

Thus, the Einstein field equation becomes

$$R_{\mu\nu} - \frac{1}{2} R g_{\mu\nu} - \Lambda g_{\mu\nu} = -\frac{8\pi G}{c^4} T_{\mu\nu}$$

[23] For a derivation of the Einstein tensor from the Riemann and Ricci tensors, see http://www7b.biglobe.ne.jp/~kcy05t/rieman.html.

[24] Recall that the covariant derivative gives the "absolute" derivative of a vector or tensor, that is, that which is not due to change in the basis vectors. But the metric Tensor is defined entirely in terms of basis vectors, so that its covariant derivative is always zero.

where Λ is called the cosmological constant.[25]

Application to Cosmology[26]

At relatively small scales, the density and pressure in the universe undergo strong variations, and hence so does the curvature of spacetime. But not so at large scales. Moreover, by the cosmological principle, the large-scale structure of the universe must look the same from any point in it. Thus, on a large scale, the curvature of the universe must be the same everywhere. As Gauss showed, there are only three surfaces on which the curvature is the same at any point: the plane, the sphere and a hard to visualize object called the pseudosphere. These surfaces have constant curvature κ/Z^2, where Z is the radius and where κ is 0 for the plane, 1 for the sphere and -1 for the pseudosphere. But since we are now talking of the curvature of space itself, we must now think of these surfaces as three-dimensional rather than two-dimensional ones. The metric on any such surface is

$$ds^2 = \frac{dr^2}{1-\kappa r^2 / Z^2} + r^2 d\Omega^2,$$

where r is distance from the origin and $d\Omega$ is an angular displacement.

Spacetime interval in a space of constant curvature is given by

$$ds^2 = Z^2\left[\frac{d\rho^2}{1-\kappa\rho^2} + \rho^2 d\Omega^2\right] - d\tau^2,$$

where $\rho = r/Z$ and τ is time multiplied by c. This is known as the Robertson-Walker metric.

---

[25] For an introduction to the Einstein field equation, see Mould (1994), pp. 312–321 and pp. 392–394.

[26] This section is based on Mould (1994), ch. 14, *q. v.* for a more thorough discussion.

The Russian physicist Alexander Friedmann undertook the tedious task of calculating the Christoffel symbols from this metric and applying them to the Einstein Field Equations, which resulted in two equations:

$$2Z\ddot{Z} + \dot{Z}^2 + \kappa c^2 - c^2 \Lambda Z^2 = -kc^2 pZ^2$$
$$3\left(\dot{Z}^2 + \kappa c^2\right) - c^2 \Lambda Z^2 = kc^4 \mu Z^2,$$

where p is pressure, μ is density and $k = 8\pi G/c^4$. Z was termed the "radius" of a curved surface, but since cosmological models take into account the possibility of curved surfaces of infinite extension, it is better viewed as a scale factor or stretch factor, describing how distances between comoving objects in spacetime change over time.

Standard cosmology is based, to a large extent, on Friedmann's solutions of the Einstein field equations.

Bibliography

Albert, D. Z. *Quantum Mechanics and Experience*, Cambridge, MA and London: Harvard University Press (1992).

Aristotle, *De Interpretatione*, translated by E. M. Edghill, In Ross, W. D. (ed.) *The Works of Aristotle*, vol. I, Oxford University Press (1963a).

———. *Ethica Nicomachea*, translated by W. D. Ross, In Ross, W. D. (ed.) *The Works of Aristotle*, vol. IX, Oxford University Press (1963b).

———. *Physica*, translated by R. P. Hardie and R. K. Gaye, In Ross, W. D. (ed.) *The Works of Aristotle*, vol. II, Oxford University Press (1962).

Augustine of Hippo, *Confessiones*. http://www.augustinus.it/latino/confessioni/index2.htm.

———. *De Civitate Dei*. http://www.augustinus.it/latino/cdd/index2.htm.

Baez, J., review of *The Physical Basis of the Direction of Time* by H. D. Zeh (1993). http://math.ucr.edu/home/baez/time/time.html, accessed Nov. 2014.

Bergström, L., Goobar, A. *Cosmology and Particle Physics*, Wiley (1999).

Boethius, *Consolatio Philosophiae*, edited by R. Peiper, Leipzig (1871).

Bostrom, N. 'Are you living in a computer simulation?', *Philosophical Quarterly* (2003), vol. 53, no. 211, 243–255.

Broad, C. D. 'Examination of McTaggart's Philosophy', In Van Inwagen, P., Zimmerman, D. W. (eds.) *Metaphysics: The Big Questions*, Oxford: Blackwell (1998), pp. 74–79.

D. Saudek, *Change, the Arrow of Time, and Divine Eternity in Light of Relativity Theory*, Palgrave Frontiers in Philosophy of Religion,
https://doi.org/10.1007/978-3-030-38411-1

Callender, C. 'Finding "real" time in quantum mechanics' (2007). http://philsci-archive.pitt.edu/4262/.

———. *What Makes Time Special?* Oxford University Press (2017).

Čapek, M. 'Time in relativity theory: Arguments for a philosophy of becoming', In Fraser, J. T. (ed.) *The Voices of Time*, 2nd edn, Amherst, MA: University of Massachusetts Press (1981), pp. 434–454.

Cheng, T. P. *Relativity, Gravitation, and Cosmology*, 2nd edn, Oxford University Press (2010).

Clark, R. W. *Einstein: The Life and Times*, London: Hodder (1979).

Clarke, W. N. 'A new look at the immutability of God', In Roth, R. J. (ed.) *God Knowable and Unknowable*, New York: Fordham University Press (1973), pp. 43–72.

———. 'Is a natural theology still possible today?', In Russell, R. J., Stoeger, W. R., Coyne, G. V. (eds.) *Physics, Philosophy, and Theology*, Vatican Observatory (1988), pp. 103–123.

Craig, W. L. *The Tensed Theory of Time: A Critical Examination*, Dordrecht: Kluwer (2000a).

———. *The Tenseless Theory of Time: A Critical Examination*, Dordrecht: Kluwer (2000b).

———. *God, Time, and Eternity*, Dordrecht: Kluwer (2001a).

———. *Time and the Metaphysics of Relativity*, Dordrecht: Kluwer (2001b).

Crisp, T. M. 'Presentism', In Loux, M. J., Zimmerman D. W. (eds.) *The Oxford Handbook of Metaphysics*, Oxford University Press (2003), pp. 211–245.

Curiel, E. 'The constraints general relativity places on physicalist accounts of causality'. http://strangebeautiful.com/papers/curiel-gr-constraints-phys-caus.pdf. First published in *Theoria*, 15(1) (2000), 33–58. Pagination follows that of the online version.

Dainton, B. *Time and Space*, 2nd edn, Durham: Acumen (2010).

Davies, P. C. W. *The Physics of Time Asymmetry*, University of California Press (1977).

———. *The Goldilocks Enigma*, London: Penguin Books (2006).

De Beauregard, O. C. 'La grandeur physique temps', In Piaget, J. (ed.) *Logique et Connaissance Scientifique*, Éditions Gallimard (1967), pp. 726–753.

———. 'Time in relativity theory: Arguments for a philosophy of being', In Fraser, J. T. (ed.) *The Voices of Time*, 2nd edn, Amherst, MA: University of Massachusetts Press (1981), pp. 417–433.

De Laplace, P. S. *Théorie Analytique des Probabilités*, Paris: Courcier (1814).

D'Inverno, R. *Introducing Einstein's Relativity*, Oxford University Press (1992).

Dorato, M. *Time and Reality: Spacetime Physics and the Objectivity of Temporal Becoming*, Cooperativa Libraria Universitaria Editrice Bologna (1995).

———. 'On becoming, cosmic time and rotating universes', In Callender, C. (ed.) *Time, Reality and Experience*, Cambridge University Press (2002), pp. 253–276.

Ducheyne, S. 'Isaac Newton on space and time: metaphysician or not?', *Philosophica* (2001) 67(1), 77–114.

Eddington, A. *Space, Time and Gravitation: An Outline of the General Relativity Theory*, Cambridge University Press (1920).

Ellis, G. F. R. 'Issues in the philosophy of cosmology' (2008). http://arxiv.org/pdf/astro-ph/0602280v2.pdf.

Ellis, G. F. R., Williams, R. M. *Flat and Curved Space-Times*, 2nd edn, Oxford University Press (2000).

Fleisch, D. *A Student's Guide to Vectors and Tensors*, Cambridge University Press (2012).

Frisch, M. 'Time and causation', In Bardon, A., Dyke, H. (eds.) *A Companion to the Philosophy of Time*, John Wiley & Sons (2013), pp. 282–300. http://faculty.philosophy.umd.edu/mfrisch/papers/time-and-causation.pdf.

Gödel, K. 'An example of a new type of cosmological solutions of Einstein's field equations of gravitation' (1949a), In Feferman, S., et al. (eds.) *Kurt Gödel – Collected Works*, vol. II, Oxford University Press (1990), pp. 190–198.

———. 'A remark about the relationship between relativity theory and idealistic philosophy' (1949b), In Feferman, S., et al. (eds.) *Kurt Gödel – Collected Works*, vol. II, Oxford University Press (1990), pp. 202–207.

Grinfeld, P. *Introduction to Tensor Analysis and the Calculus of Moving Surfaces*, New York: Springer (2013).

Grünbaum, A. *Philosophical Problems of Space and Time*, 2nd edn, Dordrecht & Boston: D. Reidel (1973).

Harrington, J. 'Special relativity and the future: A defense of the point present', *Studies in History and Philosophy of Science Part B: Studies in History and Philosophy of Modern Physics* (2008) 39(1), 82–101. Preprint version: http://philsci-archive.pitt.edu/2139/1/TimeSurvey.pdf. Pagination follows that of the preprint version.

Hartle, J. B. 'The physics of "now"' (2008). https://arxiv.org/abs/gr-qc/0403001. First published in *American Journal of Physics*, (2005) 73, 101–109. Pagination follows that of the online version.

Hausdorff, F. *Grundzüge der Mengenlehre*, New York: Chelsea (1978).

Hawking, S. W. *A Stubbornly Persistent Illusion: The Essential Scientific Works of Albert Einstein*, Philadelphia: Running Press (2009).

———, and Ellis, G. F. R. *The Large Scale Structure of Space-Time*, Cambridge University Press (1973).

Holmes, R. *Elementary Set Theory with a Universal Set*, Cahiers du Centre de logique (2012). http://math.boisestate.edu/~holmes/holmes/head.pdf.

Hume, D. *Enquiry Concerning Human Understanding*, edited by J. McCormack, M. W. Calkins, Leipzig: Meiner (1913).

———. *A Treatise of Human Nature*, edited by L. A. Selby-Bigge, P. H. Nidditch, 2nd edn, Oxford University Press (1978).

Isham, C. J., Polkinghorne, J. C. 'The debate over the block universe', In Russell, R. J., Murphy, N., Isham, C. J. (eds.) *Quantum Cosmology and the Laws of Nature: Scientific Perspectives on Divine Action*, 2nd edn, Vatican and Berkeley (1999), pp. 139–147.

Janich, P. *Die Protophysik der Zeit*, 1st edn. Frankfurt am Main: Suhrkamp (1980).

———. *Das Maß der Dinge: Protophysik von Raum, Zeit und Materie*, 2nd edn, Frankfurt am Main: Suhrkamp (2016).

Kant, I. *Kritik der reinen Vernunft* (2nd edn, 1787), In Preußische Akademie der Wissenschaften (ed.) *Kant's gesammelte Schriften* (1900ff.), vol. III.

———. *Critique of Pure Reason*, translated by J. M. D. Meiklejohn, London: Henry G. Bohn (1855).

Kretzmann, N. 'Omniscience and immutability', *The Journal of Philosophy* (1966) 63(14), 409–421.

Kroes, P. *Time: Its Structure and Role in Physical Theories*, Dordrecht: Reidel (1985).

Lieber, L. R. *The Einstein Theory of Relativity*, edited by D. Derbes and R. Jantzen, Philadelphia: Paul Dry Books (2008).

Löffler, W. *Einführung in die Logik*, Stuttgart: Kohlhammer (2008).

———. *Einführung in die Religionsphilosophie*, 2nd edn, Darmstadt: Wissenschaftliche Buchgesellschaft (2013).

Lowe, E. J. 'Individuation', In Loux, M. J., Zimmerman, D. W. (eds.) *The Oxford Handbook of Metaphysics*, Oxford University Press (2003), pp. 75–95.

Lucas, J. *The Future: An Essay on God, Temporality and Truth*, Oxford: Basil Blackwell (1989).

———. 'Transcendental Tense' (1998). http://users.ox.ac.uk/~jrlucas/time/transten.html.

Margenau, H. *The Nature of Physical Reality*, New York: McGraw-Hill Book Company (1950).

McTaggart, J. M. E. 'The unreality of time', In Le Poidevin, R., MacBeath, M. (eds.) *The Philosophy of Time*, Oxford University Press (1993), pp. 23–34.

Mellor, D. H. *Real Time II*, London and New York: Routledge (1998).

Moling, M. *Zeit und Ewigkeit nach Thomas von Aquin*, PhD thesis at Innsbruck University (2005).

Morrison, M. A. *Understanding Quantum Physics: A User's Manual*, New Jersey: Prentice Hall (1990).

Mould, R. A. *Basic Relativity*, New York: Springer (1994).

Mumford, S., Anjum, R. L. *Getting Causes from Powers*, Oxford University Press (2011).

Newton, I. *Philosophiae Naturalis Principia Mathematica*, Geneva: Barrillot et Filii (1739).

Padgett, A. *God, Eternity, and the Nature of Time*, Basingstoke: Macmillan (1992).

Pannenberg, W. *Systematische Theologie*, vols. I. (1988) and II. (1991), Göttingen: Vandenhoeck & Ruprecht.

——— 'Eternity, time, and space', *Zygon* (2005) 40(1), 97–106.

Pinnock, C., Rice, R., Sanders, J., Hasker, W., Basinger, D. *The Openness of God: A Biblical Challenge to the Traditional Understanding of God*, Downers Grove: InterVarsity Press (1994).

Plantinga, A. *Warranted Christian Belief*, Oxford University Press (2000).

Plato, *The Republic*, translated by J. L. Davies and D. J. Vaughan, London and New York: Macmillan (1892).

———. *Timaeus*, translated by B. Jowett. https://www.ellopos.net/elpenor/physis/plato-timaeus/default.asp.

Popper, K. *The Open Universe: An Argument for Indeterminism*, Cambridge University Press (1988).

Prantl, K. *Aristoteles' Werke*, vol. I (1978), Aalen: Scientia Verlag.

Price, H. *Time's Arrow and Archimedes' Point: New Directions for the Physics of Time*, Oxford University Press (1996).

———, and Corry, R. 'A case for causal republicanism?', In Price, H., Corry, R. (eds.) *Causation, Physics, and the Constitution of Reality: Russell's Republic Revisited*, Oxford University Press (2007), pp. 1–10.

Putnam, H. 'Time and physical geometry', *The Journal of Philosophy* (1967) 64(8), 240–247.

Reichenbach, H. *The Philosophy of Space and Time*, New York: Dover Publications (1958).

———. *Philosophie der Raum-Zeit-Lehre*, edited by A. Kamlah and M. Reichenbach, Braunschweig: Vieweg (1977).

Rietdijk, C. W. 'A rigorous proof of determinism derived from the special theory of relativity', *Philosophy of Science* (1966) 33(4), 341–344.

Roark, T. *Aristotle on Time: A Study of the Physics*, Cambridge University Press (2011).

Robinson, H. 'Substance'. In Zalta, E. N. (ed.) *The Stanford Encyclopedia of Philosophy* (Spring 2014 Edition). http://plato.stanford.edu/archives/spr2014/entries/substance/.

Rucker, R. *Infinity and the Mind*, Brighton: Harvester Press (1982).

Rugh, S. E., Zinkernagel, H. 'On the physical basis of cosmic time' (date unknown). http://core.kmi.open.ac.uk/download/pdf/11921921.pdf.

———. 'A critical note on time in the multiverse' (2013). https://arxiv.org/abs/1305.2055.

Runggaldier, E., Kanzian, C. *Grundprobleme der Analytischen Ontologie*, Stuttgart: UTB (1998).

Russell, R. J. *Time in Eternity: Pannenberg, Physics, and Eschatology in Creative Mutual Interaction*, University of Notre Dame Press (2012).

Saudek D. 'Beyond A-theory and the block universe: a non-circular derivation of "before", change, and the local arrow of time', *Kriterion* (2019). http://www.kriterion-journal-of-philosophy.org/kriterion/issues/Permanent/Kriterion-saudek-01.pdf.

Savitt, S. 'Time in the special theory of relativity', In Callender, C., Arntzenius, F. (eds.) The Oxford Handbook of Philosophy of Time, Oxford University Press (2011), pp. 546–570.

Schilpp, P. A. *The Philosophy of Rudolf Carnap*, Cambridge University Press (1963).

Schutz, B. F. *A First Course in General Relativity*, 2nd edn, Cambridge University Press (2009).

Schütz, L. *Thomas-Lexikon*, 2nd edn, Paderborn: Schöningh (1895).

Schwarz, G. *Raum und Zeit als naturphilosophisches Problem*, 2nd edn, Wien: WUV – Universitätsverlag (1992).

Silberstein, M., Stuckey, W. M., McDevitt, T. *Beyond the Dynamical Universe: Unifying Block Universe Physics and Time as Experienced*, Oxford University Press (2018).

Stein, H. 'On Einstein-Minkowski space-time', *The Journal of Philosophy* (1968) 65(1), 5–23.

———. 'On relativity theory and openness of the future', *Philosophy of Science* (1991) 58(2), 147–167.

Strumia, A. *The Problem of Foundations: An Adventurous Navigation from Sets to Entities, from Gödel to Thomas Aquinas*, Siena: Cantagalli (2012).

Suárez, F. *Disputationes Metaphysicae*, edited by C. Berton, reprinted at Hildesheim by G. Olms Verlagsbuchhandlung (1965).

Taylor, E. F., Wheeler, J. A. *Spacetime Physics*, 2nd edn, New York: W. H. Freeman and Company (1992).

Thomas Aquinas, *Summa Theologiae*, edited by P. Caramello, Torino: Marietti (1962).

Tooley, M. *Time, Tense, and Causation*, Oxford: Clarendon Press (1997).

Unger, R. M., Smolin, L. *The Singular Universe and the Reality of Time: A Proposal in Natural Philosophy*, Cambridge University Press (2015).

Von Weizsäcker, C. F. *Aufbau der Physik*, Carl Hanser Verlag (2002).

Weyl, H. *Philosophie der Mathematik und der Naturwissenschaft*, 3rd edn, München: R. Oldenbourg (1966).

Wheeler, J. A., Feynman, R. P. 'Interaction with the absorber as the mechanism of radiation', *Reviews of Modern Physics* (1945) 17(2–3), 157–161.

Whitrow, G. J. *The Natural Philosophy of Time*, London and Edinburgh: Thomas Nelson and Sons (1963).

Wittgenstein, L. *Tractatus logico-philosophicus*, In *Ludwig Wittgenstein Werkausgabe*, 2nd edn, vol. I (1995), Frankfurt am Main: Suhrkamp.

Youhei, T. 'Derivation of Riemann, Ricci, Einstein tensors'. http://www7b.biglobe.ne.jp/~kcy05t/rieman.html.

Young, H. D., Freedman, R. A. *University Physics*, 13th edn, Addison-Wesley (2012).

Yourgrau, P. *The Disappearance of Time: Kurt Gödel and the Idealistic Tradition in Philosophy*, Cambridge University Press (1991).

Zeh, H. D. *The Physical Basis of the Direction of Time*, 3rd edn, Springer (1999).

Zimmerman, D. W. 'Presentism and the space-time manifold', In Callender, C. (ed.) *The Oxford Handbook of Philosophy of Time*, Oxford University Press (2013), pp. 163–246. http://fas-philosophy.rutgers.edu/zimmerman/Presentism%20and%20Rel.for.Web.2.pdf. Pagination follows that of the online version.

Zimmerman, E. J. 'Time and quantum theory', In Fraser, J. T. (ed.) *The Voices of Time* (1981), 2nd edn, Amherst, MA: University of Massachusetts Press, pp. 479–499.

Author Index[1]

[1] Note: Page numbers followed by 'n' refer to notes.

D. Saudek, *Change, the Arrow of Time, and Divine Eternity in Light of Relativity Theory*, Palgrave Frontiers in Philosophy of Religion,
https://doi.org/10.1007/978-3-030-38411-1

Subject Index[1]

[1] Note: Page numbers followed by 'n' refer to notes.

D. Saudek, *Change, the Arrow of Time, and Divine Eternity in Light of Relativity Theory*, Palgrave Frontiers in Philosophy of Religion,
https://doi.org/10.1007/978-3-030-38411-1

The manufacturer's authorised representative in the EU is Springer Nature Customer Service Centre GmbH, Europaplatz 3, 69115 Heidelberg, Germany. If you have any concerns regarding our products, please contact ProductSafety@springernature.com

Printed and bound by CPI Group (UK) Ltd, Croydon, CR0 4YY
22/09/2026
02221054-0003